Quar

Jim Al-Khalili, OBE, is a leading theoretical physicist and senior lecturer at the University of Surrey in Guildford. He is also a regular broadcaster of television and radio documentaries, including the BAFTA-nominated *Chemistry: A Volatile History* and *The Secret Life of Chaos*, and he presents *The Life Scientific* on BBC Radio 4. A Fellow of the UK Institute of Physics, he was awarded the Royal Society Michael Faraday medal in 2007 and the Institute of Physics Kelvin medal in 2011. Jim Al-Khalili lives with his wife and two children in Southsea in Hampshire.

Quantum

A Guide for the Perplexed

Jim Al-Khalili

WEIDENFELD & NICOLSON

A W&N PAPERBACK

First published in Great Britain in 2003
by Weidenfeld & Nicolson
This paperback edition published in 2012
by Weidenfeld & Nicolson
an imprint of Orion Books Ltd,
Carmelite House, 50 Victoria Embankment,
London EC4Y 0DZ

An Hachette UK company

10

A CIP catalogue record for this book
is available from the British Library.

ISBN 978-1-7802-2395-7

Printed in Great Britain by Clays Ltd, St Ives plc

The Orion Publishing Group's policy is to use papers that
are natural, renewable and recyclable products and made
from wood grown in sustainable forests. The logging and
manufacturing processes are expected to conform to
the environmental regulations of the country of origin.

www.orionbooks.co.uk

Contents

This book is dedicated to my father, to whom I am indebted for, among many things, first telling me about some strange theory called quantum mechanics.

Introduction

During my teens, I was an avid reader of a magazine called *The Unexplained* that was full of accounts of UFO sightings, Bermuda Triangle stories, and assorted other paranormal phenomena. I remember the tingle of excitement I would feel as I opened each issue to be reassured that the world was full of weird and wonderful occurrences that no one understood. Best of all were the fascinating photographs that seemed to have been taken with a cheap camera, a shaky hand, in thick fog on a dark night. They would purport to provide evidence for flying saucers, ghostly apparitions, and Loch Ness monsters. I particularly remember one morbid picture showing the charred remains of an old lady's dismembered feet, still inside their cosy slippers and lying next to a pile of ash in a living room; all that remained of the old dear after an incident of 'spontaneous human combustion'.

I have no idea whether that magazine is still going – I certainly have not come across it recently – but the public's fascination with all manner of paranormal phenomena that appear to have avoided being neatly labelled, classified and packaged by science continues unabated. It seems many people take comfort in the knowledge that there are still corners of our world that are holding out against science's inexorable advance, where magic, mystery, and other-worldliness still survive and thrive.

This is quite a shame; I find it frustrating that all science's victories in explaining and rationalizing the multitude of phenomena in our Universe are sometimes regarded as mundane or lacking in wonder. One physicist who let this get under his skin was Richard Feynman, who won the Nobel Prize in 1965 for his contribution to our understanding of the nature of light. He wrote:

'Poets say science takes away from the beauty of stars – mere globs of gas atoms. Nothing is "mere". I too see the stars on a desert night, and feel them. But do I see less or more? ... What is the pattern, or the meaning, or the why? It does not do harm to the mystery to know a little more about it. For far more marvellous is the truth than any artists of the past imagined it. Why do the poets of the present not speak of it?'

These days, with so many popularizations of science that the public has access to in books, magazines, television documentaries, and the Internet, I believe attitudes are changing. But there remains one area of science that cannot be entirely rationalized using everyday language, or explained in simple, easily digestible concepts and sound bites. I refer not to any speculative, half-baked idea based on some pseudo-scientific arguments such as ESP or, worse still, astrology. On the contrary, this subject is very much mainstream science. In fact, it is a field of study that is so pervasive, so fundamental to our understanding of nature, that it underpins a large fraction of all physical sciences. It is described by a theory the discovery of which was without doubt the single most important scientific advance of the twentieth century. By some curious coincidence, it is also the subject of this book.

Quantum mechanics is remarkable for two seemingly contradictory reasons. On the one hand, it is so fundamental to our understanding of the workings of our world that it lies at the very heart of most of the technological advances made in the past half a century. On the other hand, no one seems to know exactly what it means!

When it comes to the world of the quantum we really are crossing into a quite extraordinary domain. A domain where it seems we are free to choose any one of a number of explanations for what is observed, each of which is in its way so astonishingly strange that it even makes tales of alien abductions sound perfectly reasonable.

If only people knew how frustratingly and yet wonderfully un-mundane the quantum world really is, how our familiar and

solid reality ultimately rests so tenuously on an unfathomable ghostly reality beneath. No need any longer for tales of the Bermuda Triangle or poltergeist activities; quantum phenomena are much stranger. And while just about every recorded paranormal incident can be explained away with no more than a pinch of common sense, quantum theory has been tested, prodded, and probed in every imaginable way for nearly a hundred years. It is a pity none of the predictions of quantum mechanics made it, as far as I am aware, into an issue of *The Unexplained*.

I must make it clear from the outset that it is not the theory of quantum mechanics that is weird or illogical. On the contrary, it is a beautifully accurate and logical mathematical construction that describes Nature superbly well. In fact, without quantum mechanics we would not be able to understand the basics of modern chemistry, or electronics, or material science. Without quantum mechanics we would not have invented the silicon chip or the laser; there would be no television sets, computers, microwaves, CD and DVD players, mobile phones, and so much more that we take for granted in our technological age.

Quantum mechanics accurately predicts and explains the behaviour of the very building blocks of matter – not just the atoms, but the particles that make up the atoms – with incredible accuracy. It has led us to a very precise and almost complete understanding of how subatomic particles interact with each other and connect up to form the world we see around us, and of which we are of course a part.

Thus, we seem to be faced with a bit of a contradiction. How can a scientific theory be so successful in explaining so many 'how's and 'why's, and yet still be so obscure?

Most practising physicists who use the rules and mathematical formulae of quantum mechanics on a daily basis will say that they do not have a problem with it. After all, they know that it works. It has helped us to understand a vast array of phenomena in nature, its mathematical framework and formulation is precise and well understood and, despite the numerous attempts of many who

have doubted it, has survived with flying colours every conceivable experimental test thrown at it. Indeed, it is not uncommon for physicists to become irritated by those of their colleagues who still feel unable to come to terms with the counter-intuitive and bizarre nature of the subatomic world forced upon us by the theory. After all, what right do we have to expect nature at the unimaginably tiny scale of atoms to behave in a way familiar to us from our everyday experiences on the scale of humans, cars, trees, and buildings? It is not that the theory of quantum mechanics is a strange description of Nature but that Nature herself behaves in a surprising and counter-intuitive way. And if quantum mechanics provides us with the theoretical tools to understand everything we observe then we have no right to blame Nature – or the theory – for our intellectual shortcomings.

Many physicists become impatient with those seeking a more intuitive interpretation of quantum mechanics, and take what is in my view a rather unscientific stance. They will say 'Why don't you just shut up and simply use the quantum tools to make predictions about results of experiments? It is a futile waste of time to seek complete enlightenment about anything that cannot be checked experimentally.'

In fact, the standard interpretation of quantum mechanics – the one that is in general taught to all physics students – has built into it strict rules and conditions that physicists must adhere to regarding the sort of information they are able to extract from Nature given a particular experimental set-up. I know this must sound unnecessarily obscure to appear so early on in the book but you must appreciate from the outset that quantum mechanics is like no other intellectual human endeavour, either before it or since.

Like most physicists I have spent many years thinking about quantum mechanics, both from a professional point of view as a practising researcher and as someone interested in its deeper meaning – a field known as the foundations of quantum mechanics. Maybe the twenty years or so I have been grappling with quantum

mechanics is not enough time for me to have 'come to terms' with it yet. But I feel I have heard enough sides of the debate (and believe me it is still ongoing despite the optimistic and in some ways dishonest claims to the contrary by those adhering to a particular interpretation) for me to at least take a step back from the fray. Most of what I cover in this book will, I hope, not be controversial, and where I do touch on issues relating to 'what is going on' I hope to adopt a neutral and objective position. I am not a supporter of any particular interpretation of quantum mechanics, but I do have clear views on the issue. You are of course free to disagree with these but I am sure I can win you over – unless you are one of the 'shut up and calculate' brigade, in which case you should not be reading this book but doing something more useful instead!

All I will say for the time being is that my favoured version is called the 'shut up while you calculate' interpretation. This way I am free to worry about quantum mechanics when I am not busy using it.

But this book is not just about the meaning of quantum mechanics. It is also about its successes, both in explaining so many phenomena, and in its many past, present, and future applications in our everyday lives. I will thus take you on a journey through philosophy, subatomic physics, and theories of higher dimensions to today's high-tech world of lasers and microchips and tomorrow's remarkable world of quantum magic.

But while I hope this all sounds fascinating, it is natural for the complete novice to the field to first ask what all the fuss is about. There are many ways of highlighting the weird nature of quantum mechanics, some from everyday examples we are familiar with and which we take for granted, others by employing 'thought experiments': idealized situations that do not need to be realized in the laboratory to be appreciated. Indeed, nothing brings home so ruthlessly and beautifully the mystery of quantum mechanics as the experiment with the double slit. So, that is where I will begin.

Chapter 1
Nature's
Conjuring Trick

Before I start throwing around too much science this early on in the book I will describe a simple experiment. It will, I predict, sound like magic. Indeed you may well wish not to believe a word of it; that is up to you. Like any magician worth his salt, I will not, at this stage, reveal to you exactly how or why it works. Unlike a conjuring trick, however, you will slowly begin to appreciate as the story unfolds that there is no sleight of hand, no hidden mirrors or secret compartments. In fact, you should be left concluding that there is no rational explanation of how things could possibly be the way I outline them.

Since I can only use adjectives such as 'weird', 'strange' and 'mysterious' just so many times, I will waste no further time with this fanfare and get on with it. What I will describe is a real experiment and you will have to trust me that what is seen is not just theoretical speculation. The experiment is simple to do given the right apparatus and has been performed many times in many different ways. It is also important to point out that I shall describe the experiment, not using the benefit of an understanding of quantum physics, but from the point of view of the reader who does not yet know what to expect or how to come to terms with the astonishing results. I will assume that you will be trying to rationalize the results logically as we go along according to what you might regard as common sense, which is quite

different to the way a quantum physicist would explain things. That will come later.

I should first say that the trick, if I may still refer to it as a trick for now, could be performed simply by shining light on a special screen; and indeed this is often the way it is described in many texts. However, it turns out that the nature of light is itself very strange, and this reduces the dramatic effect. We learn at school that light behaves as a wave; it can be made up of different wavelengths (which give the different colours of the spectrum we see in a rainbow). It exhibits all the properties we expect of waves, such as interference (when two waves mix), diffraction (the spreading out of waves when they are squeezed through a narrow gap), and refraction (the bending of a wave as it travels through different transparent media). These phenomena are to do with the way waves behave when they encounter a barrier or when two waves meet. The reason I mention that light is strange is because this wave-like behaviour is not the whole story. In fact, Einstein won his Nobel prize for showing that light can sometimes exhibit some very unwave-like behaviour; but more of that in the next chapter. For the purposes of the two-slit trick, we can assume that light is a wave, since this doesn't ruin the really good part.

First, a beam of light is shone on a screen with two narrow slits in it that allow some of the light to pass through to a second screen where an interference pattern is seen. This is a sequence of light and dark bands that are due to the way the separate light waves emerging from the two slits spread out, overlap and merge before hitting the back screen. Where two wave crests (or troughs) meet they combine together to form a higher crest (or lower trough) that corresponds to more intense light and hence a bright band on the screen. But where a crest of one wave corresponds with a trough of the other they cancel out resulting in a dark patch. In between these two extremes some light survives and there is a gradual blending in of the pattern on the screen. It is therefore only because light behaves as a wave washing through both slits simultaneously that the interference pattern appears. No problem so far I hope.

Light shone through two narrow slits will form a pattern of fringes on the screen due to interference between the light waves emerging from the slits. This will of course only happen if the light source is 'monochromatic' (consisting of light of a single wavelength).

Next, a similar experiment is carried out using sand. This time the second screen is placed below the one with the slits and gravity does the work. As the sand falls onto the first screen, separate piles gradually build up on the lower one beneath the two slits. This is not surprising since each individual grain of sand must pass through one or other of the two slits; we are not dealing with waves now and there is no interference. The two piles of sand will be of the same height provided the two slits are of the same size and the sand is poured from a position above their mid-point.

Now for the interesting part: repeating the trick with atoms. A special apparatus – let us call it an atomic gun for want of a better name – fires a beam of atoms at a screen with two appropriately narrow slits.[1] On the other side, the second screen is treated with

[1] The slits would indeed need to be narrow and very close together for this. Such experiments were in fact carried out in the 1990s in which the screen was a sheet of gold foil and the slits were of the order of a single micrometre (a thousandth of a millimetre) wide.

Grains of sand do not behave as waves of course and form two piles beneath the slits.

a coating that shows up a tiny bright spot wherever a single atom hits it.

Of course I don't need to tell you that atoms are incredibly tiny entities and so should clearly behave in a manner similar to the sand, as opposed to spread-out waves, capable of overlapping both slits at once.

First, we run the experiment with just one slit open. Not surprisingly, we get a spread of light spots on the back screen behind the open slit. This slight spreading of the spots might be worrying if you already know something about wave behaviour since that is what happens to a wave passing through a narrow slit (diffraction). However, we can quickly reassure ourselves that we needn't be too concerned just yet as some of the atoms may just be bumping off the edges of the slit rather than going cleanly through and this might account for the spread.

Next, we open the second slit and wait for the spots to appear on the screen. If I asked you to now predict the distribution of the bright spots that builds up you would naturally guess that it would look like the two piles of sand. Namely, that a cluster of spots builds up behind each slit, giving two distinct patches of light that are

Now repeat the trick with atoms. When one of the slits is closed the atoms only pass through the open slit. The distribution of spots indicates where the atoms have landed. While this slight spreading is actually due to a wave property called diffraction, we can still argue that the atoms are behaving as particles and the result is no different to one of the sand piles.

brightest in their centre and gradually fade away as we move out and the 'hits' become rarer. The mid-point between the two bright patches will be dark, corresponding as it does to a region of the screen that is equally hard to reach for the atoms whichever slit they manage to get through.

Well, surprise, surprise, atoms just don't behave like this. Instead, we see an interference pattern of light and dark fringes just as we did with light. The brightest part of the screen, believe it or not, is in the centre where we would not expect many atoms to be able to reach!

We could have a stab at explaining how the pattern might be forming in the following way. Despite an atom being a tiny localized particle (after all, each atom hits the screen at a single point) it seems that the stream of atoms have somehow conspired to behave in a way similar to a wave. They wash up against the first screen, and those that manage to get through the slits 'interfere' with each other's paths, via atomic forces, in a way that mimics exactly the pattern that is produced when the peaks and troughs of two waves come together.

Maybe the atoms bump into each other in a particular coordinated way such as to guide each other onto the screen. Atoms, we would reason, are most certainly not like spread-out waves (such as light, or water waves, or sound waves); but then maybe we should not expect them to behave exactly like grains of sand either.

So here is where this comforting prop is knocked away. To begin with, we see that the pattern of fringes on the back screen is connected somehow to the way two waves interfere. Just as with normal waves, its details depend on the width of the slits, the distance between them and how far away the back screen is.

This in itself is not proof that the atoms are behaving in a wave-like fashion. However, not only has the double-slit experiment been performed with atoms, but it has also been done by firing individual atoms one at a time! That is, only when we see the flash of light on the back screen signalling the arrival of the atom do we fire the next one, and so on. There is only ever one atom travelling through our apparatus at any one time. Each atom that manages to get through the slits leaves a tiny localized spot of light somewhere on the screen. In practice, most of the atoms are stopped by the first screen rather than pass through its narrow slits, and so we are only interested in those that do get through.

What we see is quite incredible. The spots gradually build up on the screen and light bands of an interference pattern slowly emerge where there is a high density of spots. In between these bands are dark regions where no or very few atoms land.

It seems we can no longer argue that atoms emerging from one slit bump into atoms emerging from the other. The interference pattern cannot be the result of a collective behaviour. So what is going on? What makes this result particularly spectacular is that there are places on the back screen where atoms were arriving when only one of the slits was open. By opening the second slit we are providing another route for the atoms to go through, so you would expect to increase the chances of atoms reaching these places. Instead, with both slits open no atoms ever arrive at all. Somehow, if the atom really does only go through one slit then it must already

know whether or not the other one is open, and act accordingly!

To recap, each atom fired from the gun leaves it as a tiny 'localized' particle and arrives at the second screen also as a particle, as is evident from the tiny flash of light when it arrives. But in between, as it encounters the two slits, there is something mysterious going on, akin to the behaviour of a spread-out wave that gets split into two components, each emerging from a slit and interfering with the other on the other side. How else can we rationalize the way the atom has to be aware of both slits at the same time?

When I used to perform conjuring tricks at my kids' birthday parties – and they are now too old for such embarrassments – there were always a few smart alecs who would announce that they know how the tricks are performed. They would insist on looking up my sleeves and behind the screen and under the table to catch me out. Such normally annoying behaviour is positively encouraged in scientific experiments. So let us try to look up nature's sleeve by lying in wait behind one of the two slits to see what the atoms actually do. This can be achieved by setting up an atom detector behind one of the slits so that it can catch any atom passing through that slit. We find that an atom is registered every now and then. We never catch part of an atom. At least that would prove that the 'rest of the atom' had gone through the other slit. Sometimes of course an atom will go through the other slit, as witnessed by a spot of light appearing on the screen. Naturally, the accumulation of many spots on the screen does not now have the feature of an interference pattern since atoms are only getting through one of the slits, just as they did in the first part of the experiment when only one slit was open. Now, instead of closing the second slit we have caught all the atoms that go through it in our detector.

You should now be beginning to doubt the truth of what I am saying. It is one thing for atoms to magically transform themselves from tiny particles into spread-out waves whenever they encounter two possible routes through the first screen. Maybe there is an as yet unexplained physical process that takes place. But it is another matter entirely to suggest that the atom can somehow be aware of

Left: With both slits open, the atoms are fired through one at a time. Only after we see a spot appearing on the screen do we send the next one through. Each atom seems to land at a random spot on the screen and, to begin with, there is no obvious pattern. Gradually, as the number of spots builds up, an interference pattern of bands emerges. What is going on? How can the atoms conspire to form this pattern that is a result of wave-like behaviour? It seems that each atom is more likely to land in certain regions than in others. Clearly, some wave-like process is involved in the propagation of a single atom. But the interference pattern only arises when a wave goes through *both* *slits*. How does a tiny atom, which leaves the gun as a localized particle and hits the screen at a definite point, go through both slits at once?

the detector hiding behind one of the slits ready to catch it in the act of its spread-out state. It is as though it knows beforehand that we are lying in wait ready to ambush it and cunningly maintains its particle persona!

But even here we have not really added anything new to the original set-up. Presumably the detector somehow has the ability to convert a spread-out 'wave' atom back to a localized particle just as the back screen does whenever an atom reaches it.

The detector can be set up in a less intrusive way so as to be able to simply register a 'signal' as an atom passes through that slit on its way to the screen. If an atom is not detected but a hit is recorded on the back screen then it must have gone through the other slit.[2] Of course, I am over-simplifying here; we will see later that the detector cannot register a signal without being very intrusive.

So, you might think that we have proof at last that the atoms do indeed go through either one slit or the other, as we have every right to expect, and not simultaneously through both like a spread-out wave. But before you get too smug take a look at the screen. Once enough atoms have registered a signal in the detector as they pass through the slit that is under surveillance and you are thus convinced that half went through one slit and half through the other, you will find that the interference pattern has disappeared! In its place are just two bright patches due to the collection of a pile of atoms behind

[2] I am assuming here that the detector has 100 per cent efficiency and will definitely fire if an atom passes through the slit it is watching.

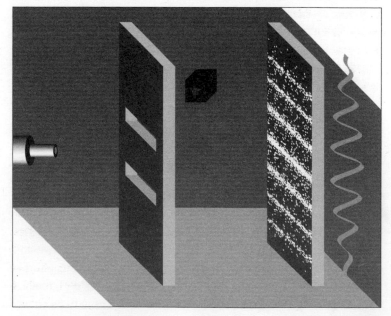

Left: With a detector in place that records which slit each atom passes through, the interference pattern disappears. It is as though the atoms do not wish to be caught in the act of going both ways at once, and only travel through one slit or the other. Two bands form on the screen adjacent to the slits as a result of particle-like behaviour, similar to what happens with the sand.

Below left: With the detector turned off we now have no knowledge of the route taken by each atom. Now that their secret is safe, the atoms revert to their mysterious wave-like behaviour and the interference pattern comes back!

each slit. The atoms are now behaving like particles, just like the grains of sand. It is as though each atom behaves like a wave when it is confronted by the slits, unless we are spying on it in which case it innocently remains as a tiny particle. Crazy, isn't it?

Of course you might be the kind of person who is very hard to please and even now think this not too surprising. Maybe the mere presence of a large detector in the path of the atoms might somehow upset its strange and delicate behaviour. But it seems this is not the problem, for switching the detector off – and therefore having no knowledge of which slit the atom has passed through – allows the interference pattern to build up again. It is only when the atom is being watched that it remains as a particle throughout. Clearly the act of observing the atom is crucial.

As if all this is not enough, there is one final twist to the trick. Even if we admit that atoms are crafty little things, maybe they are not crafty enough! How about if we let the atoms get through the slits, one at a time of course, and allowing them to do whatever it is that atoms do in order to give the interference pattern on the back screen. But this time we make sure we catch them in the act. In what are known as 'delayed choice' experiments it is possible to have a detector in place and only switch it on after the atom has gone through the slits. We can be sure of this by controlling the energy of the fired atoms and thus knowing how long it would take for any atom to reach the first screen.

This sort of experiment has indeed been carried out using photons rather than atoms, but the argument remains the same. With modern high-speed electronics the detector can be close enough to one of the

slits to be able to tell whether the atom had come through it, and yet it need only be switched on after the atom, behaving like a spread-out wave, has emerged from both the slits, but before it reaches the detector. Surely now it is too late for the atom to suddenly decide to behave like a localized particle that has only passed through one of the slits. Apparently not. In such experiments, the interference pattern is nevertheless found to disappear.

What is going on? This seems like magic, and I suspect you probably do not believe me. Well, physicists have spent many years trying to come up with a logical explanation for what is seen. Here is where I must be careful to qualify what I mean by a 'logical explanation'. I am using it in the loose, everyday sense, meaning an explanation that sits comfortably within the bounds of what we might regard as rational, reasonable and sensible, and not in contradiction or conflict with the behaviour of the other phenomena of which we have more direct experience.

In fact, quantum mechanics does provide us with a perfectly logical explanation of the two-slit trick. But it is an explanation only of what we observe and not of what is going on when we are not looking. But since all we have to go on is what we can see and measure, maybe it makes no sense to ask for more. How can we assess the legitimacy or truth of an account of a phenomenon that we can never, even in principle, check? As soon as we try, we alter the outcome.

Maybe I am asking too much of the word 'logical'. After all, there are many instances in everyday life where we might regard the behaviour of something as illogical or irrational. All this means is that this behaviour was unexpected in some sense. Eventually, we should in principle be able to analyse the behaviour based on the notion of cause and effect; that this happens and therefore that happens as a consequence and so on. It does not matter how complex the chain of events are that lead to a certain behaviour, or even that we can fully understand each step. What matters is that, somehow, what is observed can be explained. There may be new processes at work, new forces or properties of nature that have not yet been understood or

even discovered. All that matters is that we can use logic, however convoluted, to explain what might possibly be going on.

Physicists have been forced to admit that, in the case of the double-slit trick, there is no rational way out. We can explain what we see but not why. However strange you may find the predictions of quantum mechanics, it must be emphasized that it is not the theory – mankind's invention – that is strange, but rather Nature herself that insists on such a strange kind of reality on the microscopic scale.

A few years ago I read that Robert Frost's poem *The Road Not Taken* had been voted by Americans as their most popular poem of all time. Frost, long regarded as America's best-loved 20th-century poet, spent most of his life in New England where he wrote mainly about the rural life in the surrounding countryside of New Hampshire. The somewhat melancholic *The Road Not Taken* is a beautiful example of this. It also happens to touch – quite unintentionally on the part of Frost – on the very essence of what the quantum world must be like:

Two roads diverged in a yellow wood,
And sorry I could not travel both
And be one traveller, long I stood
And looked down one as long as I could
To where it bent in the undergrowth;

Then took the other as just as fair,
And having perhaps the better claim,
Because it was grassy and wanted wear;
Though as for that the passing there
Had worn them really about the same,

And both that morning equally lay
In leaves no step had trodden black.
Oh, I kept the first for another day!
Yet knowing how way leads to way,
I doubted if I should ever come back.

I shall be telling this with a sigh
Somewhere ages and ages hence:
Two roads diverged in a wood, and I –
I took the one less travelled by,
And that has made all the difference.

While we are often burdened with regrets about the choices we make in life, quantum mechanics tells of a very different reality at the subatomic level. Meeting it for the first time, the quantum world may seem unbelievable to us when judged according to

The quantum skier. To highlight just how strange the behaviour of quantum particles really is, it would be as though a skier, faced with having to go round a tree blocking his path, decided instead to go both ways at once. Clearly, this would be regarded, in our everyday world of trees and skiers, as some kind of hoax. But it really does happen in the quantum world.

the prejudiced views of our everyday experiences – what we call common sense. But the alien way that quantum objects behave is beyond any doubt. A single atom can travel down both roads in Frost's yellow wood... no regrets for atoms; they can sample all possible experiences simultaneously. Indeed, they follow the advice of the great American baseball player, Yogi Berra, who said, 'If you come to a fork in the road, take it.'

What we have seen in this chapter is just one example of the way in which the quantum phenomenon known as 'superposition' manifests itself. I could have described any one of a number of equally baffling 'tricks' that rely on quantum superposition, along with several other fascinating features unique to the quantum domain. I hope this chapter has not put you off continuing on the exciting journey ahead.

Buckyballs and the Dual-Slit Experiment

Professor Markus Arndt & Professor Anton Zeilinger
Department of Physics, University of Vienna

We usually think of a physical body as a localized object while the notion of a wave is intimately linked to something extended and delocalized. Contrary to this common belief quantum physics claims that both, seemingly contradictory, notions can apply to one and the same object in one and the same experiment.

We have recently implemented such an experiment with large carbon molecules called buckyballs. These molecules, known as C_{60} and C_{70}, contain sixty or seventy carbon atoms each, arranged to form the smallest known replica of a soccer ball, with a diameter no bigger than one millionth of a millimetre. In spite of their small size these molecules are the most massive objects ever used to demonstrate the wave-like nature of matter to date.

The experiment is set up as follows. The molecule source is a simple oven, filled with the carbon powder. The molecules can escape from a hole, like water vapour escaping from a hot kettle. They then fly through two collimating slits towards a laser detector with high resolution that can be shifted to record the spatial distribution of the molecular beam.

On the way towards the detector the molecules may encounter three different possibilities – either no obstacle at all, or a very narrow slit or a very fine grating, which is a membrane with several slits.

The molecular beam profile for the first, 'empty', case is a single narrow peak and is in complete agreement with our naïve expectation, assuming that each molecule can be regarded as a free-flying classical ball.

However, the first weirdness occurs in the second case: If we place a single, very narrow slit

– 70 nanometres (millionths of a millimetre) wide – in between the source and the detector we find a profile on the screen that differs from the empty case. We notice a strong broadening – instead of the narrowing, which we would have expected if the molecules were just little soccer balls. This is a consequence of diffraction, a property of waves.

The situation becomes even stranger when we replace the narrow slit by a grating. This structure is now composed of several openings, slightly narrower (nominally 50 nanometres) than the first slit. The slits are regularly spaced (about 50 nanometres apart). If molecules were simple particles we would expect an increased signal everywhere on the screen. But – to the surprise of our common sense – we now find that there are positions where we hardly detect any molecules at all.

Opening two or more pathways in the wall, instead of only one, reduces the number of detected molecules at certain places. This is very counter-intuitive and can no longer be explained with the model of classical balls flying along well defined paths, but it is in perfect agreement with a model based on the wave-nature of single molecules. Here we give up on the concept of a 'trajectory' and allow the molecules to simultaneously explore an extended space, which is orders of magnitude larger than the molecule itself, resulting in quantum interference.

It is important to note that the clicks in the detector are well localized and that the location where an individual molecule arrives is absolutely random, as far as we can tell. But still, the weird wave-like pattern is built up as more and more molecules hit the detector.

Chapter 2
Origins

Many books on popular science, even physics textbooks, tend to promulgate two myths concerning the origin of quantum mechanics. Of course, oversimplified accounts of the development of science are quite common, indeed necessary, in its teaching. Most of scientific progress is a messy and slow process and it is only with hindsight, and with an overall understanding of a theory or a phenomenon, that its story can be told pedagogically instead of chronologically. This necessitates the distilling of certain events and personalities from the mêlée, often due to the neat packaging that results from the dishing out of Nobel prizes.

So what are these two myths?

The first is the oversimplified and inaccurate account of the state of physics at the end of the 19th century. The story goes that the scientists of the time felt they had most of physics wrapped up and explained; that all physical phenomena could be completely understood within a worldview supported by the twin pillars of Isaac Newton's mechanics and laws of motion and James Maxwell's newly completed electromagnetic theory. There just remained a handful of 'i's to dot and 't's to cross.

The second myth is that the German physicist Max Planck proposed a revolutionary new formula to describe an experimental result in the field of thermodynamics[1] that could not be reproduced

[1] From 'thermo', meaning 'heat', and 'dynamics', meaning motion. This field of science is to do with how heat, and other forms of energy, are transferred between bodies.

by the prevailing theory, and right away the quantum revolution was born.

How it all began

While this book is not meant to be about the history of quantum mechanics or the personalities involved in its development, I will nevertheless tell in this chapter the story of how and why it all came about. Thus, while I do not wish to dwell for too long on the state of physics before quantum mechanics, it is interesting to try and pinpoint exactly when and how it all began. Regarding the first myth, the truth is that by the time the 19th century came to a close, there were so many issues to resolve and strange phenomena to explain that clearly something had to give. Physicists and chemists could not even agree on whether matter was ultimately composed of indivisible atoms or whether it was continuous and infinitely divisible. Nor could they decide on whether or not Newton's mechanics (equations that governed how macroscopic objects[2] interacted and moved under the influence of forces) could be cast in terms of Maxwell's more fundamental theory of electromagnetism.

And as if such fundamental questions were not enough, relatively new fields of physics, such as thermodynamics and statistical mechanics,[3] were generating heated debate. On the experimental side, there were the unexplained phenomena of the photoelectric effect and black body radiation (both of which I will describe soon), and no one understood how to interpret the meaning of the patterns of 'line spectra' in the light given off by certain elements. To add to all this, there was worldwide excitement generated by the newly

[2] Basically, macroscopic means everything we can see around us that is large enough to behave in a sensible 'non-quantum' way, while microscopic simply means having a size (at the level of atoms and below) small enough for the quantum rules to govern its behaviour.

[3] The study of how to derive macroscopic properties of matter from microscopic physics.

discovered mysterious phenomena of X-rays (1895) and radioactivity (1896), not to mention the electron (1897). Basically, physics was in a glorious mess.

The second myth is that in late 1900 Max Planck revolutionized science by proposing that energy came in lumps (called 'quanta') – a notion he was forced to introduce in order to understand how warm objects radiated their heat – and quantum theory was instantly up and running. In truth, it was far less clear-cut than that. Indeed, some historians of science deny that Planck deserves any credit at all for 'discovering' quantum theory.[4] Unlike many other great revolutions in science, quantum mechanics was not due to the stroke of genius of one man. Newton had his eureka moment while contemplating an apple falling from a tree on his mother's farm, and hit upon his famous Law of Gravitation (although it is likely that this event was apocryphal). And no one will deny credit to Darwin for his theory of evolution, nor to Einstein for his theories of relativity. But the discovery of quantum mechanics would have been simply too much for one person. Its development took thirty years and the combined intellectual might of the world's greatest minds.

Before I proceed, here is a good place to explain why I interchange between 'quantum theory' and 'quantum mechanics'. The former is used to refer to the state of affairs during the period 1900–1920 when everything was at the level of simple postulates and formulae that helped clarify some issues surrounding the nature of light and the structure of atoms. It wasn't until the 1920s that the real revolution took place, and a completely new worldview (quantum mechanics) replaced Newton's 'mechanics' when it came to describing the underlying structure of the subatomic world.

But back to the question of how it all got started, and let us be fair about this. Planck was awarded the Nobel Prize for physics in 1918

[4]Two good sources for a discussion of this controversy are the works of the philosopher Thomas Kuhn in *Black-Body Theory and the Quantum Discontinuity: 1894–1912* (Clarendon Press, Oxford, 1978), and the historian of science Helge Kragh in *Quantum Generations: A History of Physics in the Twentieth Century* (Princeton University Press, 1999).

Planck's constant

According to Planck's formula, the energy of the smallest bundle of light of a given frequency (a single quantum) is equal to the frequency multiplied by a certain constant. This is known as Planck's constant of action. It has the symbol h and, like the speed of light c, is one of the universal constants of nature.

The relation between energy and frequency is very simple. For instance, the frequency of violet light at one extreme of the visible spectrum is twice that of red light at the other end, and so a quantum of violet light has twice as much energy as a quantum of red light.

Today, every student of physics learns about Planck's constant. In units of kilograms, metres and seconds it has the incredibly tiny value of 6.63×10^{-34}, and yet it is one of the most important numbers in science. Despite this number being so small, the crucial point is that it is not zero, otherwise there would be no quantum behaviour.

Very often, Planck's constant is combined with another fundamental constant of nature, pi (π). This number, as every school kid is taught, is the ratio of a circle's circumference to its diameter and crops up all the time in physics equations. In fact, the quantity $h/2\pi$ appears so often in quantum mechanics that a new symbol has been invented to describe it, called HBAR (pronounced 'aich-bar').

with the citation: 'in recognition of the services he rendered to the advancement of physics by his discovery of energy quanta'. So while we will see how others such as Einstein and Boltzmann can also lay claim to laying the foundations of the original quantum theory, the key to all this is, after all, the concept of the 'quantum'; something that was first introduced in Planck's simple formula. So what exactly did he do?

Planck grew up in Munich and studied in Berlin, obtaining his PhD when he was only 21. Ten years later he was a full professor of physics. But it was to be a further eleven years before he proposed his famous formula in a lecture to the Berlin Physical Society. He had come up with it in an ad hoc manner in order to solve a long-standing problem to do with the way some objects radiate heat. But he regarded it more as a neat mathematical trick than as containing a deep truth about nature itself. [5]

Black-body radiation

The Sun's heat, or thermal radiation, that you feel on your face on a summer's day has travelled through the vacuum of space to reach us. What you might not be aware of is that this radiation has covered the distance between the Sun and the Earth in exactly the same time (about eight minutes) that it has taken for the Sun's light to reach us. The reason for this is that both the Sun's thermal and visible radiation are forms of electromagnetic waves. All that distinguishes them from each other is their wavelength. The oscillations corresponding to visible light are squeezed more closely together (shorter wavelengths – and thus higher frequency) than those of the waves we feel as heat. The Sun also emits even shorter wavelength ultraviolet light that is beyond the visible spectrum.

But it is not just the Sun that emits electromagnetic radiation. All

[5] Planck later found it hard to accept the predictions of quantum theory and spent many years trying to find a way round its conclusions.

bodies do, and over the whole frequency range of the spectrum. The distribution over frequency depends on the body's temperature. If a solid is hot enough it will glow visibly, but as it cools its glow will diminish as the longer wavelength radiation – beyond the visible – dominates. This does not mean that it ceases to emit visible light, but simply that the intensity of the light will be too weak for us to see. Of course all matter also absorbs and reflects radiation falling on it. Which wavelengths are absorbed and reflected defines the colour of everything we see.

Physicists in the second half of the 19th century were very interested in the way a particular type of warm object, known as a black body, emits radiation. Black bodies are so called because they are perfect absorbers of radiation and do not reflect any light or heat. Of course, a black body must somehow dispose of all the energy it absorbs – otherwise its temperature would become infinite! Therefore, it radiates off its heat over all possible wavelengths. The wavelength of the most intense radiation depends of course on the temperature of the black body.

In almost all physics textbooks you will find a graph showing several curves (called spectra) produced when the intensity of radiation emitted by a black body is plotted against the radiation's wavelength,[6] at various temperatures. These curves all start off at low intensity for the very short wavelengths, rise to a maximum and fall off again at longer wavelengths. It was the exact shape of these curves that so interested physicists such as Max Planck.

It is often the way in scientific research that once new experimental data become available it is the job of theories to explain it. So it was with the black-body spectra. In 1896, a colleague of Planck's, Wilhelm Wien, came up with a formula that allowed him to plot a curve that was in very good agreement with his own precisely measured experimental data down at the short wavelength end, but

[6] Sometimes the plot is against frequency rather than wavelength. However, since these two properties of waves are equivalent (short wavelengths means high frequency and long wavelengths low frequency) the information from the two types of plots is the same.

which didn't agree very well at the longer wavelengths.

At around the same time, one of the giants of 19th-century physics, the Englishman Lord Rayleigh, proposed a different formula based on a more rigorous theoretical derivation than Wien's equation. However, his theory suffered from the opposite problem: it matched the data perfectly well on the longer wavelength side of the curve, but broke down completely on the other side: at radiation wavelengths shorter than those of visible light. This failure of Rayleigh's theory manifested itself in a curve which predicted that the thermal radiation emitted by a black body should increase in intensity as the wavelength got shorter, and shoot up to infinity in the ultraviolet region of the spectrum. This problem became known as the 'ultraviolet catastrophe'.

Contrary to many popular accounts, Max Planck became interested in black-body radiation not because of the failure of Rayleigh's formula,[7] but in order to place Wien's formula on a firm theoretical foundation. After his early attempts failed, there followed a period of very intense work after which he tentatively, and rather reluctantly, arrived at a new and quite different formula.

Planck was conservative in his views and, in the early part of his career, did not even believe in the existence of atoms, as advocated by contemporaries of his such as Ludwig Boltzmann. Planck felt that it would soon be proved that matter was continuous in the sense that it was not ultimately composed of fundamental 'building blocks', but could be infinitely divided up and still retain its essence. However, in finding a solution to the black-body radiation problem, it was Boltzmann's ideas that he based his theory on. He presented his results in a seminar to the German Physical Society on 14 December 1900, a date widely regarded as the birthday of quantum physics.

His suggestion was this: if a black body is ultimately composed of vibrating atoms – although it should be stressed that Planck simply referred to them as 'oscillators' (vague elementary entities that oscillate, or vibrate, at a frequency that depends on the body's

[7] The term 'ultraviolet catastrophe' was not even used until 1911.

temperature) – then the energy they give off (the black body's radiation) depends on their frequency of vibration. This would mean that the higher their frequency, the more energy they would emit. But the key point was that such oscillators would only have certain modes of vibration and their frequencies would have to ramp up in definite steps rather than smoothly.[8] Therefore, the energy given off could only have certain values since not all possible energies would be allowed. That is, the energy would come in discrete lumps, or 'quanta'. This was a radical departure from Maxwell's electromagnetic theory, in which energy is regarded as continuous.

Two things need to be mentioned here. Firstly, Planck did not at first realize the implications of his revolutionary idea. His introduction of energy quanta was, in his own words, 'a purely formal assumption and I really did not give it much thought except that, no matter what the cost, I must bring about a positive result'. Secondly, Planck did not regard all energy as being ultimately composed of irreducible tiny lumps. That had to wait five more years for the genius of Einstein.

To recap then, Planck's hypothesis was based on two assumptions: the first was that the energy of the atoms (or oscillators) can only take on certain values. These are simple multiples of the vibrating frequency of the atoms. The second was that emission of radiation by a black body is associated with the energy of the atoms dropping from one value, or level, to a lower one. When the energy drops, the atom emits a single quantum of radiation energy.

The easiest way to visualize this is to think of a ball rolling down a flight of stairs and hence giving up its 'potential' energy in jumps

[8] It is a bit like the difference between the notes played on a guitar and a violin. A guitar has 'frets' – metal bridges – on its neck. When a string is pressed onto the neck below a fret with your finger the note that is played is a result of the string's vibration between the fret and the bridge at the other end. The different lengths of vibrating string resulting from pressing down behind two successive frets cause a difference in pitch of half a step. So, just like a piano, a guitar cannot (without certain skill) make a sound in between two half steps. Whereas a violin does not have frets so its strings can have any frequency, or pitch, you want depending on exactly where you place your finger.

instead of continuously, as when it rolls down a smooth slope. The difference is that the quantum jumps between the atomic energy levels happen instantaneously, whereas the ball's potential energy does in fact pass through all energy levels since it takes a short but finite time to drop down each step.

The importance of Planck's work was not appreciated right away. In the words of the historian Helge Kragh:

'If a revolution occurred in physics in December 1900, nobody seemed to notice it. Planck was no exception, and the importance ascribed to his work is largely an historical reconstruction.'

This seems rather harsh, but probably true. I would prefer to be more generous and put it slightly differently. Planck is rightly the founding father of the quantum; he just didn't know it at the time! It took other, deeper and more original thinkers to really appreciate what he had started. In any case, Planck's contribution was just the first small step. Physicists such as Einstein, Bohr, de Broglie, Schrödinger, and Heisenberg contributed, individually, more than Planck. It is just that Planck was the first.

I have always had a soft spot for Max Planck, who had a difficult life. As an elder statesman of science he resisted the Nazis in the 1930s, but great personal tragedy was to befall him during the Second World War. He had decided to remain in Germany during the Nazi regime where he openly opposed many of their policies, particularly the persecution of the Jews. Three of his children had already died at a young age, and his two remaining sons did not survive the war. One was killed in action and the other executed for his part in a failed attempt to assassinate Hitler. Planck himself suffered great hardship when allied bombs destroyed his house in 1944. He died in 1947 aged 89.

Einstein

Had Einstein not discovered relativity theory, he would no doubt have still become a household name, partly for his role in the development of quantum theory. But since he stands head and shoulders above every other physicist, apart from Isaac Newton, it just seems unfair on all the rest that he is given credit for both of the great revolutions in 20th-century science (relativity and quantum theory).

In 1905, while Einstein was only 26 and working as a clerk in the Swiss Patent office, he published five theoretical papers in physics journals. Three of these papers were so important that any one of them would have ensured him a place in history.

The most famous, and indeed the most important, was the last of the five. It was his paper on special relativity, in which he showed that another fundamental tenet of Newtonian physics, the notion that space and time were absolute, was an illusion. He started with two simple postulates. The first was that the laws of nature remain the same no matter how fast you are moving, and so no one can claim to be truly at rest and all motion is relative. The second was that the speed of light through empty space is a fundamental constant of nature measured to have the same value no matter what speed the observer is travelling at. With these two ideas we are led to the conclusion that both time and space are aspects of a grander four-dimensional spacetime. Einstein also proved that the speed of light is the maximum speed possible in the Universe. Special relativity forces us to accept strange notions such as time slowing down when we travel very fast. It also leads to Einstein's best-known equation, relating mass and energy: $E=mc^2$.

Just before this paper, Einstein published another showing detailed calculations to describe Brownian motion. This is a phenomenon first observed by Scottish botanist Robert Brown in 1827, whereby pollen grains suspended in water are observed under a microscope to bounce around erratically. Einstein proved mathematically that this was due to the constant random motion of the water molecules, which was the

intensity of radiation

frequency of radiation

The old wave theory of light predicted that the higher the frequency of the radiation emitted from a black body, the greater its intensity. This intensity became infinite at ultraviolet frequencies. Something was clearly wrong with the theory.

Planck's curve

Planck's quantum theory predicted a curve that matched the old wave theory at frequencies through the visible spectrum, but did not continue to rise as the frequency increased. Instead, his theory predicted that the intensity should drop back down again to zero, in perfect agreement with experimental data.

Before Planck, it was assumed that energy radiated from a black body was continuous and that any value was possible – like a ball rolling down a smooth slope. Planck proposed that energy was quantized and as such, only certain discrete values would be allowed.

first real proof of the existence of atoms. Supporters of the idea of matter being ultimately composed of tiny indivisible units were well aware that Brownian motion might be due to the motion of atoms, but it was Einstein who confirmed this. Experiments based on his work finally convinced the last remaining opponents of the existence of atoms.

However, it is the first of Einstein's three major papers of 1905, in which he explained the origin of a phenomenon known as the 'photoelectric effect', that is of most interest to our story. Planck's formula had been pretty much ignored for five years. Einstein resurrected it and took its conclusions an important step further.

Particles of light

The photoelectric effect was another phenomenon of 19th-century physics that could not be explained at the time. It involved shining light on an electrically charged metal plate and knocking out electrons from the surface. By carefully examining how this took place, scientists found it to be in even starker contradiction with the prevailing wave theory of light than the problem of black-body radiation.

There are in fact three puzzling features of this effect. Firstly, you would think that if light has the ability to knock out particles of matter, then their energy would presumably depend on the brightness, or intensity, of the light. Surprisingly, it was found that the ability of the light to kick out the electrons depended instead on its wavelength. This is a rather unexpected result if we think of light as a wave, because increasing its intensity, and hence its energy, would imply increasing the size of its oscillations. Think of water waves crashing against the seashore; they have more energy the higher they are, not the faster they break against the shore. In the photoelectric effect, high-intensity light did not give rise to higher-energy ejected electrons, just more of them!

The second, related, puzzling feature was that, according to the wave theory, the photoelectric effect should occur at any frequency

Bright, higher intensity light is a result of a larger number of photons than dim light. But the average energy of a single photon is the same in the two cases.

of light as long as it was intense enough to provide the electrons with the necessary energy to escape. What was observed, however, is that there is a cut-off frequency below which no electrons are emitted, no matter how bright the light is.

Finally, if exposed to the energy of a light wave, the wave theory suggests that electrons would need a finite time to absorb enough energy to break away from the surface, especially if the light is feeble. No time lag was detected. Electrons were knocked out as soon as light was shone on the surface.

Einstein successfully explained this effect by extending Planck's idea of lumps of light energy. Remember that Planck had not gone so far as insisting that all radiation be quantized. Instead, all he proposed was that a black body radiated energy in packets as a direct consequence of the properties of matter. But he still believed that, in general, electromagnetic radiation was continuous. Einstein's

proposal was that all light is ultimately made up of energy quanta,[9] now known as photons. This was more than Planck had been prepared to accept.

The dual nature of light

Planck and Einstein's contribution to the quantum revolution was the first step. Looking back today with everything we know about the richness of quantum mechanics and the phenomena it can explain, we see that the idea of light being composed of particles is not such a big deal. After all, Isaac Newton himself believed that light was made up of particles, or 'corpuscles' as he called them. A contemporary of Newton's, the Dutch astronomer Christiaan Huygens, developed a rival wave theory of light. But it wasn't until the beginning of the 19th century that an Englishman by the name of Thomas Young showed beyond doubt that light had to be treated as a wave.

Young conducted the two-slit experiment[10] with light – in fact it is also known as the Young's slits experiment – and as we saw in Chapter 1, there is no mystery if we are allowed to think of waves going through both slits at once. We understand how waves do this, and it leads naturally to the interference pattern we see on the screen. It is no wonder that Young's observations banished for one hundred years the notion that light might be composed of particles. Physicists throughout the 19th century paid homage to Newton for his great achievements – and he is still rightly regarded as the greatest scientist who ever lived – but as for his idea of corpuscles of light, forget it. There was no way this picture could give rise to the interference pattern.

But then, a century after Young's experiments, Einstein proved

[9] Think of these as bundles of energy localized in space. However it took some time before the idea of light as 'particles' was truly accepted.

[10] Of course, I only called it a 'trick' in Chapter 1 to add to the dramatic effect.

Einstein's Nobel Prize

Albert Einstein was awarded the 1921 physics Nobel Prize for his explanation of the photoelectric effect, which was regarded at the time as a more significant discovery than his more famous work on relativity theory.

According to Einstein, each electron is knocked out when it is hit by a single photon of light, the energy of which depends on its frequency. He argued that the reason we do not normally see the particle-like nature of light is due to the large number of photons involved, just as we do not see individual pixels of ink on a printed image. So let us examine briefly how this picture solves

the three puzzling features of the photoelectric effect.

The first is easy. The dependence of the ejected electron's energy on the light's frequency rather than its intensity is a direct consequence of Planck's equation relating the energy of the light to its frequency.

The second feature arises because the threshold for production of the electrons only occurs when the photon energy is sufficient to release an electron. Increasing the intensity of the light just means more photons. And since photons are so minuscule and localized in space, the

The photoelectric effect involves knocking electrons off a metal surface by shining light on it. However, thinking of the light as being made up of waves does not explain observations. Only by considering it to be composed of individual particles (photons) can we explain what is seen.

likelihood that any one electron could acquire sufficient energy to escape through being hit by more than one photon is tiny.

Finally, the process is instantaneous because the electrons do not have to accumulate their energy from a wave that is spread out in space. Instead, each photon delivers all its energy to an electron in a single collision. If this energy is above the necessary threshold the electron will escape.

that light has to be regarded as a stream of particles if we are to explain the photoelectric effect!

So what is going on? It seems that we cannot think of light as purely a wave phenomenon, nor merely as composed of particles. It appears to behave like a wave under some circumstances (Young's slits) and as a collection of localized particles under others (photoelectric effect). The phenomena we have looked at so far all point to the idea that this split personality of light has to be taken seriously, despite the natural feeling of discomfort when contemplating it for the first time. Indeed, today this so-called wave-particle duality of light is not in doubt.

But wait a minute, could there be two types of light: wavy light and lumpy light? Can light alter its state depending on how we use it or detect it? Physicists do find the concept of photons rather baffling. Remember, each single photon (a particle) is associated with a definite frequency and wavelength (wave properties). So what does it mean to say that a particle has a wavelength? Spread-out waves have wavelengths; particles are, well, not spread out at all!

Bohr: physicist, philosopher, phootballer [11]

The next step in the quantum revolution was taken by a young Danish physicist by the name of Niels Bohr, who arrived in England in 1911 armed with a fresh PhD from Copenhagen and the complete works of Charles Dickens (which he used to learn English). Obviously, Bohr was not yet famous as a physicist, but he saw it as a more secure career path than football, which he also excelled at as a keen amateur. However, he was not quite at the level of his younger brother Harold who played in defence for the Danish Olympic team that lost the 1908 gold medal to Great Britain. Harold was later to become a highly respected mathematician.

The lives of Niels Bohr and this author only overlapped by two months so, unfortunately, I never got to meet him. And if I had, our conversation would not have been very productive. But I have collaborated over a number of years with someone who knew him very well. Jens Bang is a theoretical physicist who was Bohr's very last scientific assistant, and so has many stories to recount about the great man as well as a deep understanding of his philosophical views. Indeed, Bohr the philosopher is almost as famous as Bohr the scientist.

He began his quantum quest when he went to work in Manchester with the New Zealander Ernest Rutherford in 1912. Rutherford himself was at the time one of the world's leading scientists and had won the 1908 Nobel Prize for Chemistry, despite being a physicist. Bohr arrived at about the time that Rutherford came up with his model of the atom. He had just discovered that atoms are comprised of a tiny dense nucleus at the centre, surrounded by even tinier electrons.

Bohr began by trying to understand the structure of Rutherford's atomic model. Thus did he embark on half a century of endeavour and effort to understand the essence of quantum phenomena, and he

[11] And phather of quantum physics. Sorry, a pun can be pushed just so far – or is this an alliteration?

is today rightly regarded as the true father of quantum mechanics. Planck and Einstein may have started things off, but Bohr was to contribute far more.

His first success was to solve two problems associated with the structure of atoms: the origin of line spectra and the explanation of atomic stability.

Rutherford's model atom suggested that electrons exist outside the nucleus at distances thousands of times further out than the radius of the nucleus itself. This picture immediately raises the question of the stability of atoms. Firstly, physicists were certain that electrons could not be at rest within atoms since the attractive electrical force exerted by the positively charged nucleus should suck the electrons in towards it. So the simple answer would be to consider a planetary model in which the electrons are in continuous orbit around the nucleus, just as the Earth must remain in orbit around the Sun if it is to avoid plummeting towards it due to its gravitational pull.

However, Bohr was worried about a crucial difference, apart from size of course, between an atom and the Solar System. According to classical electromagnetic theory, an orbiting electron must radiate light. Consequently, it would spiral in towards the nucleus as it lost its energy. This process would happen very quickly – in about a thousandth of a billionth of a second – and atoms would collapse.

With hindsight, Bohr's idea seems obvious, but at the time it was revolutionary. He proposed that if all matter gives off radiation in lumps (as was seen with black bodies), and absorbs it in lumps (the photoelectric effect) then maybe the atoms that make up the matter simply cannot accommodate energies in between these discrete lumps.

This idea went further than Planck's postulate, where the quantization of the radiation was only due to the oscillations of the atoms in warm black bodies rather than being a feature common to all atoms due to their internal structure.

Bohr postulated that the electron energies within atoms are themselves quantized. That is, the electrons are not free to follow

Bohr's model of the hydrogen atom consisted of an electron in a fixed orbit around the atomic nucleus. If the electron were to absorb a photon of the correct frequency (middle diagram) it would gain sufficient energy to 'jump up' to a higher (further out) orbit. The atom is then said to be in an excited state. This is generally an unstable situation and the atom soon de-excites (bottom diagram). The electron spits out a photon of exactly the same energy as the first one and, in so doing, loses energy and drops back down to its 'ground state'.

any orbit they like, as would be possible according to Newton's laws of motion, but only certain 'discrete' orbits, like concentric train tracks. An electron can only drop to the next lower orbit by emitting a quantum of electromagnetic energy (a photon). Likewise, it could only jump to the next orbit out by absorbing a photon. The stability of atoms was later explained more fully by a young German genius by the name of Wolfgang Pauli, who showed that each electronic orbit could only accommodate a certain number of electrons. Thus, electrons can only jump to a lower orbit if there is room for them. We will see later that electrons cannot be thought of as tiny specks circling round the nucleus, but as spread-out waves, with each 'electron wave' stretching all the way round the nucleus.

Bohr was also able to explain the meaning of atomic spectra: the fact that elements tended to give off light at certain precise sets of

frequencies (called spectral lines) – each spectrum being unique to that element. The characteristic frequencies at which a certain type of atom could emit light corresponded to certain energies (through Planck's equation). The energies of the emitted photons correspond to the energy lost by the electrons in the atom as they drop to lower orbits.

It is worth stressing that while Bohr had applied Plank's idea of quantization to atomic structure, he couldn't explain how the electrons jumped between orbits. Just like poor old Planck, Bohr had introduced his formula in an ad hoc way. He hadn't derived it from deeper fundamental principles, as theoretical physicists tend to prefer. And while his atomic model of a miniature Solar System seemed to work very nicely, it still contained aspects from Newtonian physics, which turned out to be wrong. Worst of all, his model only seemed to work for hydrogen, which contained just the one orbiting electron! It couldn't cope with anything more complicated than that. A much more complete understanding of atomic structure required the development of full quantum mechanics ten years later.

Physicists today rightly complain that Bohr's model of the atom is still taught to schoolchildren. This is not what atoms look like.[12] The Bohr model of the hydrogen atom brought to a close the first phase of the quantum revolution, known today as the old quantum theory.

A French prince rides in

We now jump forward to the early 1920s and a young French prince by the name of Louis de Broglie,[13] who was working on his PhD. Alright, he wasn't exactly a prince (not being the eldest of his siblings) but he was a nobleman from an aristocratic family, and his ancestors had served the kings of France as far back as that other famous Louis (the XIV).

[12] You will have to wait till Chapter 7 to see a better depiction of atoms. But of course, what atoms 'look like' is not what they are really like, as we shall also see later on.

[13] Pronounced to rhyme with 'destroy'.

De Broglie submitted his thesis in 1924, in which he made a daring proposal: if light, which we find easier to consider as a wave, can, according to Planck and Einstein, sometimes behave as a stream of particles, then it would be a rather nice and symmetric feature of nature if moving particles could also sometimes behave like waves.

This might sound a little strange at first, but think of it the following way. By the 1920s physicists were pretty accustomed to Einstein's idea that matter and energy were interchangeable through his equation $E=mc^2$. This states that matter can be considered somewhat like frozen energy and the two can be transformed into each other. So, since light, or rather electromagnetic radiation, which is just a form of energy, can have a split personality, then why could matter not do likewise?

De Broglie suggested that every material object could be associated with a 'matter wave', with a wavelength that depended on its mass. The more massive a particle, the shorter the wavelength of its associated wave. Note that I use the word 'associated' here, implying that de Broglie considered material objects still as solid 'lumps' and that the wave is somehow an added extra. Whereas for light, we have seen that it is the same 'stuff', sometimes behaving as a wave and sometimes as particles.

De Broglie was inspired by the work of the American physicist Arthur Compton, who had provided further strong evidence in support of the particle nature of light. In 1923, a year before de Broglie's proposal, Compton had carried out an experiment that provided dramatic confirmation of the existence of photons. He had shone X-rays (basically high-frequency light) onto a block of graphite and found that the X-rays bounced off it at a slightly lower frequency than they had originally. This was not what was predicted by the old wave theory, in which the frequency of the light should remain unchanged. But if the X-rays were high-energy photons colliding with individual electrons in the graphite then some fraction of their energy would be lost and this, according to the Planck formula, would result in a drop in their frequency.

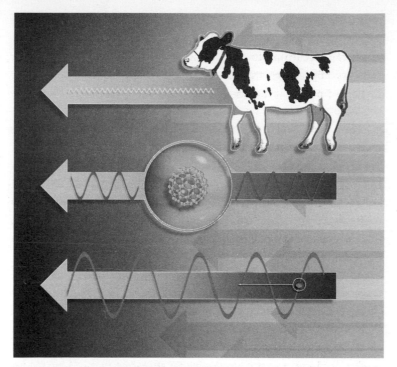

The scale of de Broglie matter waves associated with various objects.

Top: The de Broglie wavelength of a cow would be trillions of times smaller than atomic dimensions and far too tiny to ever be detected. Indeed it would be almost at a scale where the concept of space itself loses its meaning. So we need not be concerned with wavy cows.

Middle: The de Broglie wavelength of a C_{60} carbon molecule (buckyball) travelling at a few metres per second is roughly the size of the molecule itself (about a nanometre).

Bottom: An electron travelling at a few metres per second has a de Broglie wavelength equivalent to the width of a human hair (a fraction of a millimetre). This is large enough for its quantum wave nature to easily show up in experiments, and even in everyday life.

De Broglie couldn't escape the notion that there was an obvious symmetry here. Why should the photons have both wave and particle features and not the electrons? After all, Compton scattering, as the process is known today, suggests a picture of colliding solid balls. So if a photon can be treated on the same footing as an electron then

maybe the reverse might also be true. Experimental confirmation of the wave nature of electrons had to wait until 1927, when beams of electrons were first shown to give rise to interference effects – the first successful experimental confirmation of the two-slit trick with matter particles.

In the mean time, what exactly did de Broglie have in mind? There is often a good deal of confusion surrounding the issue of the wave nature of matter. In fact, de Broglie did not suggest that an electron is itself a spread out wave (though such a suggestion was indeed made by others soon after) but rather that it was a solid localized particle carried along by what is known as a wavepacket. This is an isolated piece of a wave, like a pulse, that can be constructed by superimposing many different waves of varying wavelengths and amplitudes in such a way that they interfere and cancel each other out everywhere apart from in the tiny localized region where the particle happens to be.

De Broglie came up with a formula that related the momentum of a particle, whether a photon or an electron, with the wavelength of its associated wave: the more momentum (or 'oomph') it has, the shorter its wavelength. This is why we can never detect wave-like behaviour associated with everyday objects like humans, footballs, or even grains of sand. For these objects, which are many orders of magnitude more massive than electrons, have wavelengths many orders of magnitude shorter than lengths on the subatomic scale, and so can never be detected. But could the matter waves associated with electrons, even whole atoms, be measured? Better still, if they do exist then maybe this explains the two-slit trick? Is it the atom's associated wave that goes through both the slits, while the atom itself only goes through one?

At the time, Louis de Broglie's revolutionary proposal was too radical for his contemporaries to accept. Indeed, it was touch-and-go whether he would receive his PhD, and only the last minute intervention of Einstein himself, who'd had a sneak preview of the thesis, swayed the examiners.

Soon after his work became known, events started to happen

very quickly. Physicists around Europe began to piece together the different parts of a new mathematical framework and argue over what it all meant. Not only were pieces of the jigsaw rapidly falling into place, but discoveries and insights were being made simultaneously and independently by different people, and it was only later that the connections between their ideas were made.

I will therefore end this historical chapter and move on to concentrate on what quantum mechanics tells us about the way nature behaves, rather than on how physicists made their discoveries. There are several ways that quantum mechanics can be explained, and to follow the way the founding fathers of the field developed their ideas is probably not the best one. For instance, many popularizations of quantum mechanics are outmoded in the way they describe such things as 'wave-particle duality' as being the fundamental concepts underlying the whole theory. These can often lead to muddled thinking and confusion; and that's going to be hard enough to avoid, however carefully I try.

Chapter 3
Probability
and Chance

Do you believe in fate?

For most of us this question has a quite straightforward meaning: that certain events were just meant to be, or that two people were destined to meet. But could there be any truth in the idea?

Maybe you enjoy reading your horoscope – the harmless and quite unfeasible notion that somehow the positions of the stars and planets might have an influence on how the week ahead will pan out for you. Of course while most of us do not take horoscopes seriously, the idea that the future might somehow be predictable is a very interesting one. In fact, until the quantum revolution, scientists were confident that this was indeed possible in principle, suggesting that even if we could not predict them, all future events were nevertheless somehow preordained and destined to take place.

Isaac Newton believed that every particle in the Universe should obey simple laws of motion subject to well-defined forces. This mechanistic view – one that was still shared universally by scientists and philosophers more than two centuries later – states that no matter how complex the workings of nature are, everything should be ultimately reducible to interactions between the fundamental building blocks of matter. A natural process, such as a stormy sea or the weather, may look random and unpredictable, but this is

just a consequence of its complexity and the huge number of atoms involved.

But in principle, if we could know the precise position and state of motion of every particle in a given system, no matter how many are involved, then we should be able to predict, through Newton's laws, how these particles will interact and move, and hence how the system will look at any given time in the future. In other words, a precise knowledge of the present should allow us to predict the future. This led to the Newtonian idea of a 'clockwork' universe; one in which there are, in principle, no surprises in store, since anything that can ever happen is simply a result of fundamental interactions between its parts. This is known as determinism, since the future can be completely determined if we have complete knowledge of the present.

Of course in practice such determinism is impossible for all but the simplest systems. We are well aware that forecasters cannot predict tomorrow's weather with complete confidence. We cannot even control whether a tossed coin lands 'heads' or 'tails', or where a roulette ball settles. Another current field of study in physics is chaos theory, which states that we would have to know the initial conditions of a system to infinite accuracy to be able to determine its future evolution. Chaos theory complicates the practical issues around determinism.

Indeed, simple mechanistic examples, such as the ones mentioned above, pale to insignificance when we consider how we might deal with the immense complexity of the human brain in order to understand the notion of free will. But the principle is always the same: since humans are ultimately made up of atoms too then Newton's laws should also apply in our brains. So when we make what we perceive to be a free choice about something, this is simply the mechanical processes and atomic interactions in our grey matter following deterministic laws just like everything else.

While this is a rather depressing worldview, you might feel it is OK since the idea of having sufficient information to predict the future is too incredible to entertain. But the issue is a hypothetical one: if

we had a powerful enough computer with a large enough memory to store the location and velocity of every particle in the Universe, then presumably it could calculate how the Universe would evolve.

One of the most profound changes in human thinking brought about by the quantum revolution was the notion of indeterminism – that is, the disappearance of determinism, along with the concept of the clockwork universe. So I am sorry to have to break the news to you, but 'fate' as a scientific idea was proven to be false three-quarters of a century ago.

The outcome of a game of pool

Consider how a powerful computer might attempt to model what would happen in a game of pool when the cue ball is struck into the pack. Every ball on the table will be knocked in a certain direction and most of the balls will undergo more than one collision, many bouncing off the side cushions. Of course, the computer would need to know precisely how hard the cue ball is initially struck and the exact angle it collides with the first ball in the pack. But is this enough? When all the balls have finally settled – maybe some were even potted – how close might the computer's prediction of the distribution be to reality? While it is perfectly feasible to predict the outcome when only two balls are involved in a collision, it is impossible to take into account how the complicated multiple scattering of many balls will end up. If just one ball moves at a slightly different angle then another ball, which it might have originally missed, might now glance off it and both trajectories are dramatically altered. Suddenly, the final outcome looks very different.

So, it seems we must feed into the computer more than just the initial conditions of the cue ball, but the precise arrangement of all the other balls on the table: whether they are touching each other, what the precise distances between them and the cushions are, and so on. But even this is not enough. A minute speck of dust on any

one of the balls would be enough to perturb its path by a fraction of a millimetre or cause it to slow down by a minuscule amount. Again, this will inevitably lead to a knock-on effect that alters the final arrangement. This is known as the 'butterfly effect' in chaos theory – the idea of a butterfly flapping its wings and causing a tiny change in air pressure that gradually leads to a very large change from what would have happened had the butterfly not moved, for instance causing a thunderstorm on the other side of the world some time later that would otherwise not have happened.

We would therefore need to give the computer accurate information about the precise state of the table's surface. Maybe it is more worn in some places than others. Even the temperature and humidity of the air will have a tiny effect.

Still, you can imagine that the task is not impossible. It can, in principle, be done. Of course, if there is no friction between the balls and the table then they would continue to collide and scatter for much longer, and we would therefore need to know the intitial positions of the balls to even greater accuracy to determine where they end up when they do finally stop. [1]

But 'so what?' you might be thinking. After all, since we will never be able to know everything about a particular system, we are forced to make do with assigning probabilities to different outcomes. The more we know, the more confident we can be about what might happen.

Sometimes, it is more than simply a matter of our ignorance that stops us from making a reliable prediction, but also our inability to control the initial conditions. When tossing a coin it is simply too much to ask to be able to repeat the same action and achieve the same outcome. If I toss a coin and get 'heads', it is too difficult to toss it in exactly the same way again, making it spin the same number of times so as to land 'heads' again.

This is just another way of saying we lack complete information

[1] They would gradually slow down due to air resistance and by losing energy through heat and sound when they collide.

about the system. In the pool table example, and with absolute knowledge, I could never repeat the shot by hitting the cue ball in exactly the same way to achieve an identical final outcome, with all the balls ending up in the same positions as they did the first time. Nevertheless, this repeatability is the essence of Newton's world. Such deterministic behaviour is a feature of what is known as Newtonian, or classical, mechanics. In quantum mechanics, things are very different.

Quantum unpredictability

Down in the quantum domain, we find a very serious kind of unpredictability that cannot be blamed on our ignorance of the details of the system being studied, or a practical inability to set initial conditions. Instead, it turns out to be a fundamental feature of nature itself at this level. We cannot predict with certainty what will happen next in the quantum world – not because our theories are not good enough or because we lack sufficient information, but because Nature herself operates in a very 'unpindownable' way.

It often turns out that all we can ever do in the realm of atoms is calculate probabilities for different outcomes. Such probabilities are not however assigned in the same way that we might assign probabilities to the toss of a coin or the throw of a die. Rather, quantum probabilities are built into the theory itself, and we can never, even in principle, do any better.

The radioactive decay of atomic nuclei is a good example of this; that is, identical initial situations lead to different outcomes. Consider a million identical radioactive atomic nuclei that are unstable and will, sooner or later, spontaneously 'decay' by emitting a particle and changing into a more stable form. While quantum mechanics enables us to calculate something called the half-life (the time after which half of the nuclei will have decayed) it cannot tell us when any particular nucleus will decay. The half-life is only meaningful when applied to a statistically large number of identical nuclei. We can

calculate the probability that a nucleus will have decayed after any given time, but the fact that we cannot do any better than this is not due to our ignorance.

One way out of this dilemma would simply be to say that quantum mechanics is not the whole story, and that the unpredictability of radioactive decay is indeed due to our ignorance. What we are lacking is a deeper understanding of Nature whereby we are able to predict exactly when any given nucleus might decay, just as a fuller knowledge of all the forces involved in the toss of a coin would allow us to predict its outcome. If this were true we would need to go beyond quantum mechanics to find the answer. We will see in Chapter 6 how Albert Einstein held this view; he could not accept what quantum mechanics seemed to be suggesting, that our world is, at its most fundamental level, inherently unpredictable. Indeed, one of Einstein's most famous quotes is that he did not believe 'that God plays dice', in the sense that he could not accept that Nature is probabilistic. However, Einstein was wrong.

Let us now take a closer look at the origin of quantum unpredictability and indeterminism.

Banana kicks

It is mostly thanks to Isaac Newton that we are able to understand and predict how everyday objects move and interact under the influence of forces. I remember an article in a physics magazine several years ago in which the curved trajectory of a soccer ball was analysed mathematically. The Brazilian footballer Roberto Carlos, who appeared on the front cover of the issue, is famous for his spectacular free kicks in which he is able to make a ball curl round a defensive wall far more dramatically than most footballers can manage. The trick – not that I imagine Roberto Carlos ever studied the equations in detail – was down to the precise way his kick would cause the ball to spin, and how the spinning ball then interacted with the air while

Anatomy of an equation

When we talk about 'solving' an equation for a classical particle (namely one that is not subject to quantum behaviour) we mean extracting from the algebra a value for the precise location and velocity of the particle at some future specified time. But Schrödinger's equation is different. Its solution, for the motion of an electron inside an atom say, is not just a set of numbers describing where the electron would be at any given time (as we would get if we were using Newton's equations to track the motion of the Moon around the Earth).

Instead, the solution of Schrödinger's equation is much richer. It is a mathematical quantity known as the 'wavefunction', and is symbolized by the Greek letter Ψ ('psi'). If you are looking for the origin of all quantum weirdness then you've just found it: it is all wrapped up in the wavefunction.

In elementary algebra there is always an unknown quantity 'x' – think of x as the location of a particle: 'x marks the spot' that you need to work out. In more advanced algebra the value of x may depend on what the value of a second unknown is, say t, which usually stands for 'time'. So, for example, if t=1 then maybe x=4.5, and if t=2 then x=7.3, and so on. Of course I just made these numbers up. This is the situation when we solve the equation of motion for a classical particle. Only, since the particle would exist in three-dimensional space we would need three numbers to define its location: x, y, and z. The point is that x, y, and z are just symbols that stand for certain numbers; they are not real 'things'.

The wavefunction in the Schrödinger equation is a bit like this. It is the unknown quantity, and can be worked out for each moment in time to describe the state of a quantum particle. By

'state' here I mean everything we can possibly know about the particle.

We always use mathematical symbols in physics to describe some quantity or property of the system we are studying. We use the letter 'V' for voltage, 'P' for pressure, and so on. What is different in quantum mechanics is that, unlike pressure and voltage, there is no such thing as a meter with a dial that can measure the wavefunction. While the concept of 'pressure' is a little abstract in that it is a quantity that describes the collective motion of molecules of a gas, its presence can at least be felt physically. Not so the wavefunction.

Planck's constant

Called the 'del-squared operator', this quantity describes how the wavefunction, Ψ, changes from one place to another

A mathematical quantity called an 'imaginary number'. It is equal to the square root of minus one

$$-\frac{\hbar^2}{2m}\nabla^2\Psi + V\Psi = i\hbar\frac{\partial\Psi}{\partial t}$$

The mass of the particle being described

Describes the forces acting on the particle

Describes how Ψ changes its shape with time

in flight. In a similar way, a lot of ingenuity has, over the years, gone into the design of golf balls to control their trajectories when hit in a certain way. Of course there are countless other examples. The point is that in all cases involving the motion of macroscopic bodies, the equations of motion can be solved given the necessary input data. If we know the mass and shape of a body, the precise nature of the forces acting on it and its precise current position and velocity, then we can compute, by solving the equations of motion, its precise position and velocity at all future times. This was the point behind the earlier discussion of Newtonian determinism.

Newton's equations of motion are indeed so accurate and reliable that they are able to predict the orbital motion of the planets and their moons far into the future. The equations were used by NASA to plot the trajectories of their rockets to the Moon and back. In all the above examples, specifying the current state of a physical system and the forces acting on it allows us in principle to determine precisely all future states of the system.

So why can we not apply the same equation to describe the way a microscopic particle, such as an electron, moves? If the electron is over here now, and we apply a certain force to it, for instance by switching on an electric field, then we should be able to say for definite that it will be in such and such a location five seconds from now.

Not so. It turns out that the equations that govern the behaviour of everyday objects, from grains of sand to footballs to planets, are useless in the quantum world.

The most important equation in physics

Much of the credit for the development of the theoretical understanding of quantum mechanics belongs to the Austrian physicist Erwin Schrödinger, who took the ideas of de Broglie and put them on a firm mathematical footing. It is important to note that there are several mathematical ways in which the behaviour of

a quantum system like an electron or an atom can be described, and Schrödinger's approach was just one of them. However, it is the way quantum mechanics is generally taught to undergraduate physics students and is the approach I will take here.

Schrödinger decided to see if de Broglie's idea of matter waves could somehow be used to explain Bohr's model of the atom. Remember that Bohr had suggested that electrons in atoms move around in fixed (quantized) orbits, but no one knew why this might be so. Schrödinger proposed a new equation that describes not the way a particle moves but the way a wave evolves. He came up with what is known as a wave equation.

It is now almost standard practice that popular science books devoted to ideas of modern physics will shun all mathematical equations apart from one – $E=mc^2$ – which I have already stated. But Schrödinger's equation deserves at least a quick glance (see box on page 49) if only for aesthetic reasons.[2]

Solving the Schrödinger equation provides us with a mathematical quantity called the wavefunction. This is where all the probabilistic nature of quantum mechanics comes in. For the case of an electron, for instance, the wavefunction does not give us the precise location at a moment in time, only the likelihood of the electron being found somewhere if we were to look for it there. Of course your first reaction to this should be: but this is not good enough – just telling us where the electron might be does not sound like all the information we could possibly have. Of course just stating these words doesn't actually explain anything to you. So I'll try harder.

The wavefunction contains a lot of information. At any instant in time it has a value for each point in space. So unlike the position in space of a classical particle, the wavefunction is spread out over all of space – hence the term 'wave'. But you should not think of it

[2] Whether teenagers think it cool or very sad to have it emblazoned on a T-shirt is another matter. Actually, thinking about it, I think it would be a rather nerdy thing to do.

The probability distribution of an electron trapped in a box. This is not a physical cloud describing a 'smeared out' electron but simply a mathematical cloud of probability. If we know for sure that the electron started off in one of the top corners of the box then its wavefunction will, a short time later, have spread out to occupy the whole volume of the box. However, the higher density of probability, calculated from the wavefunction, tells us that the electron is still more *likely* to be found near its original location, were we to look for it. As time goes by the probability cloud will spread out more evenly and the electron will have an equal chance of being found anywhere in the box.

as representing a real physical wave, such as a light wave. I should come clean at this point and state that no one really knows what the wavefunction actually is. Most physicists regard it as an abstract mathematical entity that can be used to extract information about nature. Others assign it to its own, very strange, independent reality. We will see in Chapter 6 how it can be possible for both views to be valid. Strangely enough, what is important here is that it doesn't matter whether the wavefunction is real or not, its mathematical properties are the same; and what it can tell us about the way nature behaves on the subatomic scale is not in any doubt.

As an example, let us consider a single electron trapped in a box. Suppose we know exactly where it is to begin with, and we feed this location into the Schrödinger equation. We will thus be able to calculate its wavefunction at some later time. Now, let us assume that we have tabulated in a computer file, or on a sheet of paper, an array of numbers that represent the values of the electron's wavefunction at various grid points around the volume of the box. We cannot use this information to locate the electron with any certainty any more. Instead, we must make do with knowing where it is most likely to be found. This is done in the following way.

At each point in space the wavefunction is assigned two numbers. The probability of the electron being in a tiny unit volume around

this point is the sum of the squares of these two numbers.[3] I tell you this simply to highlight the fact that the wavefunction itself is not a probability; you have to square it first.[4]

The probabilistic nature, and therefore the built-in unpredictability, of quantum mechanics requires a little more discussion about the nature of the wavefunction. For instance, it is helpful to give you a rough idea of how a wavefunction changes with time by considering a useful analogy.

A burglar has just been released from prison, but the local police are convinced he has not turned over a new leaf and, by studying a map of the city, can trace his likely whereabouts from the moment he is freed. While they cannot pinpoint his exact location at a given time, they can assign probabilities to burglaries being committed in various districts. To begin with, houses close to the prison are most at risk, but the area under threat grows over time. They can also say with some confidence that wealthier areas are more at risk than poorer ones. This one-man crime wave spreading through the city can be thought of as a wave of probability. It is not tangible or real, just a set of abstract numbers assigned to every part of the city. In a similar way, a wavefunction spreads out from the point where the electron was last seen and knowledge of the wavefunction allows us to assign probabilities to where it might show up next.

The police soon find out that their hunches have been accurate when a burglary is reported from some address. This alters their probability distribution since they now know the thief cannot be far away from the scene of the crime. Likewise, if the electron is detected in a certain location then its wavefunction is instantly altered. At the moment of detection there will be zero probability of finding it anywhere else. Leave it be, and its wavefunction evolves and spreads out again.

[3] Mathematically, this is because the wavefunction is what is known as a 'complex function' and therefore has what is known as both a 'real' and 'imaginary' part, but I don't need to go into that here.

[4] You never know when you might find yourself in possession of a wavefunction. It would be a shame not to know what to do with it

What is going on when we are not looking?

There is a big difference however between the example of the burglar and an electron. Even though the police can only assign probabilities to the whereabouts of the burglar, they know this is only due to their lack of complete information. After all, the burglar has not actually spread himself across the city, and while the police must regard him as being anywhere, he is really of course only ever in one place at any given time. This is so obvious it seems a waste of time telling you. But how about electrons? Most physicists believe – and with good reason, as we shall see in the next couple of chapters – that all the time we are not tracking the motion of the electron, its wavefunction is all we have at our disposal to describe it. More than that: the electron itself doesn't even exist as a simple classical particle with a definite location at each time. Its influence is spread out over space. How this can be so we can never find out. All we have is the wavefunction, and that is just a set of numbers (with physical significance of course). As soon as we look, the wavefunction is said to 'collapse' and the electron becomes a localized particle.

This should sound ridiculous and absurd to you. Why can't the electron simply be behaving as a real particle all the time? Surely just because we cannot predict with any reliability what it might be doing when we are not looking doesn't mean it isn't doing something? Well, if you are thinking along these lines, then you are not alone: Einstein also thought this. But the majority of physicists are convinced that this cosy picture is wrong. Nevertheless, a significant and growing minority of them are no longer so sure, and I will explain why in Chapter 6.

Back to our simple analogy, and the term 'crime wave'. This implies something that is undergoing oscillations: undulating peaks and troughs like ripples on the surface of a pond. Indeed, waves emanating from a point will spread out in concentric circles (as when you throw a stone into the pond). In the same way, the quantum wavefunction must somehow also be 'wavy', otherwise we would

not see the wave-like property of interference in the two-slit trick – it should be no surprise by now that the two-slit trick has something to do with the properties of wavefunctions.

In general, a wavefunction does not simply oscillate like a water wave. It is far more complicated than that. I mentioned that at each point in space the wavefunction is defined by two numbers known as its real and imaginary parts. Joining all the 'real' numbers together produces one wave and the 'imaginary' numbers another, and the full wavefunction is a combination of the two. In addition, a typical wavefunction will, if plotted on a graph, have a quite complicated shape depending on the system it is describing. A single electron trapped in a box will of course be described by a rather simple wavefunction. But the wavefunction describing the structure of an atomic nucleus, with its many protons and neutrons obeying complicated rules, is much more complicated.

Heisenberg's uncertainty principle

One of the most important consequences of the probabilistic nature of the wavefunction is the concept of indeterminacy. Do not confuse this with indeterminism, which we met earlier and which states that knowledge of certain aspects of the state a particle, such as its position at a particular moment in time, does not imply that its future position can be known with certainty. Indeterminacy on the other hand states that we can never know at the same time, and with total precision, everything about a quantum system, even if we try to measure it.

The best known example of indeterminacy is given by the uncertainty relation, first discovered by Werner Heisenberg. An electron moving about freely in space could be anywhere; we say that there is an infinite uncertainty in its position. But for an electron trapped inside a very tiny box, we can say that its location is pretty well known, and the uncertainty in its position is small. This means that the numbers associated with its wavefunction will be equal to

Heisenberg's uncertainty principle says we cannot know the precise position and momentum of a quantum particle at the same time. This 'one-aspect-or-the-other-but-not-both-simultaneously' feature of nature led Niels Bohr to his Principle of Complementarity, which states that both seemingly conflicting aspects are necessary for a complete description of a quantum particle.

The outline of this vase can also be viewed as two human profiles facing each other. However, it is difficult to see both aspects of the image at the same time; if it is two faces, there is no vase; if it's a solid vase, there are no faces.

zero everywhere apart from inside the box. Such a wavefunction is said to be localized in space.

So far, I have been describing what is referred to as the 'position wavefunction' (since we can compute from it the likelihood of finding the electron in different positions). There is another kind of quantity, called the momentum wavefunction, which tells us the probability that the electron has a certain momentum, or velocity, at any given moment. If we know the position wavefunction we can work out the momentum wavefunction, and vice versa, using a mathematical procedure known as a Fourier Transform. A localized position wavefunction will always give rise to a spread out momentum wavefunction,[5] and vice versa. So, an electron that has a localized position wavefunction, and thus a small uncertainty in its position, will always have a large uncertainty in it momentum (or velocity). Likewise, an electron whose velocity is pretty well known (through a localized momentum wavefunction) will necessarily have a spread out position wavefunction giving rise to a large uncertainty in its whereabouts.

This is the essence of Heisenberg's uncertainty principle. In its

[5] But I do not mean by this that it is spread out in space. Rather that it allows for a spread of values over a range of possible momenta.

mathematical form it says that one can never know at the same time the precise location and velocity of an electron (or any other quantum entity). This is not, as many books might tell you, as a result of the experimenter giving the electron an unavoidable random kick through the act of locating its position, and hence changing its speed and direction. Rather it is a consequence of the nature of the wavefunctions that describe the electron's possible location and state of motion even before we look.

Whether or not the electron actually has a definite location and velocity at any one time before we look is still being debated by physicists. The truth of the matter is that the uncertainty relation is a consequence of the relation between the two types of wavefunctions, and since the wavefunctions tell us everything we can ever know about the electron we simply cannot say any more about it. The uncertainty principle gives us a limit on what we can predict about a quantum state, and hence what we can know about it when we do look.

Nuclear halos

There are numerous phenomena in physics that should not be possible according to classical mechanics, and where we must appeal to Heisenberg's uncertainty principle to explain them. One such example is in my own field of research in nuclear physics. As I will explain in Chapter 7, the atomic nucleus is one of the most complex systems in physics. We are still uncovering its secrets almost a century after its discovery. It is also a place where quantum mechanics rules supreme.

Later on, we will be taking a closer look inside atomic nuclei to see how their constituent particles, the protons and neutrons, are held closely packed together by the strong nuclear force. This force acts like glue over very short ranges so its influence disappears completely beyond the surface of the nucleus.

The nuclei of the lightest elements tend to have roughly the same number of positively charged protons and electrically neutral neutrons. Nuclei that have more than the average number of either protons or neutrons for their masses tend to be unstable and transform quickly into a more stable form by converting some of the excess protons or neutrons into each other to redress the balance.

During the mid-1980s, experiments carried out by a Japanese research group at Lawrence-Berkeley Lab in California discovered a new property of very neutron-rich nuclei of the element lithium. The stable forms of lithium have atomic nuclei containing three or four neutrons bound with three protons. The Berkeley experiment discovered that the nucleus lithium-11 (three protons plus eight neutrons) seemed to have a much bigger size than expected; much more than can be accounted for by the extra neutrons. By firing a beam of such nuclei inside an accelerator at a thin carbon target they could measure how many

survived on the other side. The bigger the lithium-11 nuclei were, the more likely it would be that they would collide with the carbon nuclei and fragment. While many were expected to pass through the target unscathed, far fewer were picked up in the detectors on the other side than expected. A rough analogy would be like pouring sand through a sieve. The bigger the grains, the fewer that get through.

Theorists soon realized that they were dealing with nuclei that were like no others seen before in nature. The outer two neutrons in lithium-11 are very weakly bound to the rest of the nucleus (the 'core') and spend most of their time relatively far from this core. In fact, they float around well outside the range of the nuclear force that is holding them in place, forming what has been dubbed a 'neutron halo'. Of course the volume of the halo is still far tinier than that occupied by a lithium atom's electrons.

The neutron halo is a purely quantum phenomenon and should not exist according to classical mechanics.[6] However, I have used the same language to describe these halo neutrons as Bohr used in his old quantum theory of electrons in orbit. We know now that this is not really correct. So let me be a little more careful.

The uncertainty principle can be used to explain the large size of halo nuclei less directly. Another type of experiment used to study them is to deliberately break them up in a nuclear reaction and measure the way the pieces fly apart. It was found that the broken-up fragments stayed relatively close together as they flew off and only moved apart very slowly.

Quantum mechanically, we say that the internal momentum of the pieces has a very narrow spread about zero, or a highly localized momentum wavefunction. Since the pieces (the two neutrons and the core) are so weakly held together it doesn't take much to break up such nuclei. So the momentum wavefunction describing their relative motion after the nucleus is broken up is not so different in shape to the one describing an intact nucleus.

[6] As it is they are highly unstable with a lifetime of about a tenth of a second.

The quantum description of an atom is that of a probability cloud of electrons surrounding the tiny nucleus (top left). However, this is what we can say *before we look*. If we could take a snapshot of the atom (top right) then we would see individual electrons in specific locations. The neutron halo is also like this. It is no more than a neutron probability cloud (bottom left). The fact that it spreads out way beyond the rest of the nucleus means that, were we to somehow take a photograph of the nucleus (bottom right), we would very likely find the 'halo' neutrons some distance from all the other more tightly packed neutrons and protons in the nucleus.

The uncertainty principle tells us that this momentum wavefunction corresponds to a very spread-out position wavefunction and consequently an extended probability distribution. Thus the neutron halo is not actually two 'smeared out' neutrons, but rather a large volume of space around the core in which the neutrons have a high probability of being found. It is a cloud of neutron probability.

The golden years of quantum mechanics

The period 1925–1927 saw a revolution in quantum physics far greater than that initiated by Planck at the turn of the century. Niels Bohr had received the 1922 Nobel Prize 'for his services in the investigation of the structure of atoms, and radiation emitted from them'. But by this time a number of young European physicists were just completing their PhDs and beginning to question whether Bohr's theory of quantized electron orbits in atoms was really the last word. The three grand masters of atomic theory at the time were Bohr in Copenhagen, Arnold Sommerfeld in Munich and Max Born in Göttingen. But it was their young students who were to make the great impact.

Before 1925, physicists were aware that a major problem with Bohr's atomic theory[7] was that it was unable to explain how two electrons would interact inside atoms using the idea of Bohr's orbits. His equations worked fine for the hydrogen atom which contained just one electron, but the atomic structure of the next element up, helium, with two electrons could not be described. One of the young brat pack, Wolfgang Pauli, characterized the desperate situation in a letter to a colleague in May 1925:

'Physics at the moment is again very muddled; in any case, for me it is too complicated, and I wish I were a film comedian or something of that sort and had never heard about physics.'

The first breakthrough was to come from Werner Heisenberg, a young German who in spite

[7] Which Sommerfeld later extended and generalized to include elliptical electron orbits as well as Bohr's simple circular ones.

of his brilliance almost failed his PhD examination in 1923. While recovering from a bout of hay fever on the German island of Heligoland during the summer of 1925, he made a major advance in formulating a new mathematical theory. At the same time, Max Born and his young assistant in Göttingen, Pascual Jordan, submitted a paper in which they suggested that the 'the true laws of nature should involve only such quantities that can be observed'. As soon as Heisenberg returned to Göttingen and heard of their work he embraced it into his new theory and argued that the old Bohr-Sommerfled theory could not be correct since it relied at its heart on quantities which could never be observed, namely the electron orbits. His theory stated that only those quantities that can be measured directly, such as the electrons' energies, have any physical meaning.

By September 1925, Heisenberg, Born, and Jordan had arrived at a new theory of quantum 'mechanics'. At its heart, their idea relied on a rather strange set of mathematical relations. Basically, the product of two quantities, say A times B, was not equal to B times A. With ordinary numbers this is silly of course: obviously, 3 times 4 is the same as 4 times 3. But the quantities in their theory followed a different multiplication rule, which was already well-known to mathematicians, namely that obeyed by quantities called matrices. Very soon, this new theory became known as matrix mechanics. It was attributed to Heisenberg, but in truth the contributions of Born and Jordan should not be underestimated. The three's voluminous paper *On Quantum Mechanics II* [8] was published in February 1926.

Other young physicists such as Pauli and Dirac helped in clarifying many of the issues in the new theory and were to go on to make their own Nobel Prize-winning contributions. Heisenberg was to win the 1932 Nobel Prize.

At the same time, in January 1926, the Austrian Erwin Schrödinger submitted his first paper outlining an alternative

[8] An earlier paper, *On Quantum Mechanics*, was published three months earlier by Born and Jordan alone.

approach. His atomic theory started from de Broglie's idea of matter waves and yielded exactly the same results as that of Heisenberg. His version became known as wave mechanics. Unlike de Broglie however, Schrödinger dispensed with the idea of matter particles, such as electrons in atoms, having associated waves, and instead claimed that only the waves were real.

Schrödinger's wave mechanics and his now-famous equation were an instant success. His approach was seen by most physicists as simpler to deal with than Heisenberg's matrix formalism.

It is commonly thought that Dirac was the first person to show that the Schrödinger and Heisenberg theories were equivalent (like saying the same thing in two languages). In fact, it was Pauli who was the first to prove this – in a letter that was never published and only appeared in print many years after his death.

In the spring of 1927, Heisenberg published his famous uncertainty principle, for which he was indebted to a large extent to discussions he had with Bohr and Pauli. What is not appreciated is that the uncertainty principle relied crucially on wave mechanics. This was despite Heisenberg, along with most of the big names at the time, being strongly critical of Schrödinger's theory.

Today, it is fair to say that physics students are taught the Schrödinger approach, whereas practising theoretical physicists tend to use a combination of both matrix and wave mechanics. Schrödinger shared the 1933 Physics Nobel Prize with Paul Dirac.

Radioactive Decay

Ron Johnson Emeritus Professor of Physics, University of Surrey

Among the many successes of quantum mechanics, its account of the phenomenon of radioactivity is perhaps the most spectacular. Newton's picture of a world of particles with definite positions and velocities at all times makes a definite but spectacularly wrong prediction for the lifetime of an alpha-decaying nucleus. No matter how many times any of the protons and neutrons in the nucleus combine into an alpha particle configuration of 2 protons and 2 neutrons, Newton says the alpha particle will never get out!

The situation of an alpha particle in a radioactive nucleus can be compared to a ball rolling around at the bottom of a bowl. A ball rolling at the same speed on the table outside the bowl would have the same energy but Newton's mechanics would absolutely forbid a ball inside the bowl suddenly appearing on the table outside. To achieve this transition we would have to temporarily provide the ball with some extra energy to get it up and over the side of the bowl. Not so for a ball of the size of an alpha particle inside a bowl of the size of a nucleus.

A typical radioactive nucleus is about 0.000000000000015 metres (15 femtometres) across and an alpha particle is about 4 times smaller. Such distances are so tiny compared with the ball and the bowl that, not surprisingly, Newton's picture does not apply. In quantum mechanics the way to think about such an alpha particle is not by asking where it is and how fast it is going. You can certainly ask such questions and appropriate measurements will give answers. But a more complete picture is obtained by asking what the wavefunction looks like. To answer this question we turn to Schrödinger's equation (see box on page 49) just as we

would turn to Newton's equation in the ball-in-the-bowl case. When we explore a new part of the Universe we are not surprised to find that physicists have to use different tools.

Schrödinger's equation predicts that, for a bowl of nuclear size, the alpha particle's wavefunction may be sizeable at very large distances outside the nucleus. This is because wavefunctions have wave-like properties and are therefore not confined by the same rules as particles. The value of the wavefunction in a region of space tells us the probability of finding the alpha particle there. Hence if the Schrödinger equation predicts that the alpha particle's wavefunction is sizeable far from the nucleus then the nucleus has a chance of 'decaying'. The Schrödinger equation predicts that this will happen only if the energy is right. Because of Einstein's $E=mc^2$ equation the right energy for this to happen is when the nucleus has the right mass.

Newton's prediction for the decay probability is precisely zero even when the energy is right, which not only fails to explain observations on radioactive nuclei, but a whole host of other predictions which can be made with the alpha-particle wavefunction. For example it can be used to predict the 'half-life' of the nucleus and how this depends on its mass and how many neutrons and protons it contains. The 'half-life' is only one aspect of the alpha particle's probability amplitude. Other processes involving the interactions of alpha particles with nuclei check out other details.

The complaint is often made that quantum mechanics will not predict the instant when a particular nucleus will give off an alpha particle. When quantum mechanics gives so much detail this criticism seems like nit-picking in the light of the complete failure of the Newtonian picture to allow any decay at all!

Chapter 4
Spooky
Connections

If you felt that the discussion about wavefunctions in the last chapter was a little abstract and rather unnecessary, then maybe you need to go back and reread Chapter 1 to remind yourself what we are dealing with here. In this chapter we investigate a few more of the strange concepts forced upon us by quantum mechanics. At the root of them all is the wavefunction. Of course, I agree that it is very hard to associate the strange properties of an abstract mathematical quantity like the wavefunction with the real world. Just because we have mathematical formulae that allow us to calculate and predict properties of an atom does not mean that the wavefunction is a mathematical description of the atom itself or worse still that the wavefunction is the atom. What we do know is that whatever is going on, the quantum world is indeed very strange. But quantum mechanics describes this strangeness very accurately, as you will discover in the second half of this book.

Usually, when I try to explain physics to my wife, Julie, her eyes glaze over as she makes little effort to stifle a yawn. So when, several years ago, I had her as a captive audience I seized my opportunity. We were due to meet up with some friends at a local wine bar, but after we had arrived and ordered wine, we received the message that our friends couldn't make it. With the children booked in with their

grandparents for the night, we were in no mood to give up on our evening of freedom. So after our usual banter of how we would one day afford one of those big houses down by the seafront or, failing that, what colour we would paint the bathroom, there was a gap in the conversation.

I therefore announced that this would be an ideal occasion for me to explain the mysteries of quantum mechanics and the two-slit trick. To my astonishment Julie agreed to listen without making any sarcastic comments. An hour or so (and a bottle or two of Chilean red) later I ended with a flourish, feeling that she would finally be full of amazement at just how weird the subatomic world must be and full of admiration for her clever husband. Instead, she looked at me incredulously and slowly shook her head: 'Well, that's just a load of rubbish. I simply don't believe it!' What could I say? There are no logical steps of reasoning one can follow in explaining quantum mechanics that culminate in an unavoidable conclusion. And I could not convince her that the two-slit interference pattern is real by carrying out a demonstration there and then.

But I was neither surprised nor disappointed. After all, physicists have tried for three-quarters of a century to understand quantum mechanics and have not really succeeded. Even after many years of study I still sometimes worry about it. I understand how to use its rules and, because of my research specialization, how to apply its mathematics to study the behaviour and properties of atomic nuclei. But if asked what it all means, I am just as baffled as the next person. My advantage over you is that I am convinced there is no simple answer or straightforward and intuitive explanation.

Having discussed the probabilistic nature of quantum mechanics arising from the properties of the wavefunction in the last chapter, we now turn to some other even stranger features with appropriately intriguing-sounding names.

Superposition

The idea of superposition is not unique to quantum mechanics but is a general property of all waves.[1] Imagine watching someone dive into an empty swimming pool.[2] You will see the ripples travel outwards along the surface of the water as simple undulations all the way to the other end of the pool. This is in stark contrast to the state of the water when the pool is full of people swimming and splashing about. The turbulent shape of its surface is now due to the combined effect of many disturbances and is achieved by adding them all together. This process of adding different waves together is known as superposition. The interference pattern we saw in the two-slit experiment with light is a direct consequence of superposition of light waves emanating from the two slits. What we would like to know is whether there is a similar process at work when we fire the atoms through the slits.

So what happens when each atom encounters the screen with the slits? Maybe it gradually spreads out into a smeared out cloud that then drifts through both slits at once. But this doesn't explain

[1] Well, for the benefit of the mathematicians, I should really say that superposition is a property of all waves that are solutions of linear equations. If this statement leaves you cold ignore it; I just want to protect myself from pedantic colleagues!

[2] Empty of other swimmers that is, not empty of water.

Superposition of waves: If two stones are dropped in a pond close to each other the ripples will spread out and meet, forming a superposition that has a pattern very different to the two sets of concentric waves, due to interference.

interference. For that we need wave-like structure. Remember that all we have at our disposal – to describe the state of the atom when we are not directly observing it – is the wavefunction, which we calculate by solving Schrödinger's equation. It turns out that this equation has the same mathematical property as all other 'wave' equations in that it allows for its different solutions to be added together to give new solutions. Just as it is possible to have a superposition of water waves or light waves, it is also possible to have a superposition of wavefunctions.

Here is where we meet our next real conceptual difficulty. Consider a wavefunction that describes an electron with a particular energy. If the electron[3] is slowed down so that it has half its original energy then its wavefunction will of course be altered. However, because of the possibility of superposition of these two different wavefunctions that describe an electron with different energies, it is also possible for the electron to exist in a state that is described by a third wavefunction. This new wavefunction is (or rather has a value at each point in space that is) the sum of the first two wavefunctions: one describing a fast moving electron and another describing a slow moving electron. This means that a single electron in such a state is both fast moving and slow moving at the same time. And I don't mean that it has some average speed in between but that it has the two quite different states of motion, or energies, at once!

Even worse, an electron might be in a state described by a wavefunction that is the sum of two or more other wavefunctions, each describing an electron in a different place. So the combined wavefunction is now telling us that the electron has to be in more than one place at once! Don't worry, I sense your scepticism. After all, I keep insisting that the wavefunction is not the physical entity itself but only a mathematical description of it. In any case, we never see this strange state of affairs when we look at electrons. We only

[3] I do hope you don't mind me switching between atoms and electrons as my examples of quantum objects. It would be simpler if I just stuck with one, but then you might get the impression that quantum weirdness was unique to that object alone.

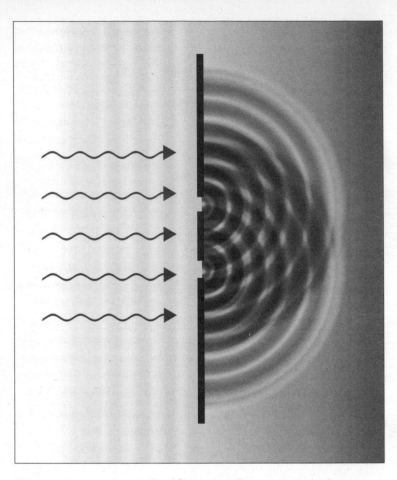

Light waves shone on the two slits of Chapter 1 are like ocean waves lapping up against a shore. On the other side, each slit acts as a new point source of light radiating semi-circular waves outwards. These will form a superposition giving rise to the interference pattern on the back screen.

ever find them in one location when we look, and only having one of their possible energies when we measure their energy. So maybe this superposition business is just a mathematical oddity after all and not a property of real particles.

For instance, it would be nice if the wavefunction just described a statistical distribution of the possible states of an electron. That is, if

The probability distribution of an electron trapped in a box. (For simplicity the box is depicted as a two-dimensional plate so that the vertical axis can be used to denote the probability density – the higher the bump, the more likely it is that the electron will be found there.)

Top: The electron is definitely somewhere in the near left corner.

Middle: The electron is definitely somewhere in the far right corner.

Bottom: The electron is in a superposition of being in two places at once. This means that if we repeatedly open the box to look for an electron described by a wavefunction with this distribution then half the time we will find it in the near left corner and the other half of the time we will find it in the far right corner. Of course we never actually see a single electron in two places at once.

we were to measure a thousand identical electrons, all described by the same superposed wavefunction, then we would find roughly half of them in the first state and the other half in the second state. Maybe we needn't worry after all; any one electron is never actually in two places at once when we look.

But wait a minute, what about the interference pattern in the two-slit trick? That was real enough, and it arose even when we sent one atom at a time through the slits! In fact, this idea of superposition of wavefunctions turns out to be the way we can explain it.

The two-slit trick 'explained'

You now know enough quantum mechanics to be able to follow what is happening in the two-slit experiment. You may not like it but then, as I have stated before, that is a quite natural reaction and, indeed, reflects your developing maturity in coming to terms with the unavoidably counter-intuitive nature of the subject. You are not meant to be comfortable with the conclusions of quantum mechanics.

Each atom fired at the slits is described by a wavefunction that evolves in time. This wavefunction is probabilistic in nature and tells us only the likely location of the atom. It is important here to stress that while we cannot regard the tiny atom as having suddenly turned into a spread-out wavefunction, the wavefunction provides us with the only means of tracking the atom from the moment it is fired till the moment it hits a particular point on the screen.

On encountering the slits, the wavefunction – being spread out

Right: How to think of the two slit trick with atoms. The atom's wavefunction is localized in space as it leaves the gun but spreads out as it travels towards the slits. In fact, the contour plots denote the probability distribution rather than the wavefunction itself, which is a more complicated quantity to visualize, even mathematically. The contours here are interpreted in the same way as you would a contour map of a hilly terrain with the inner contours denoting the highest probability of finding the atom. As the wavefunction reaches the slits it begins to squirt through both of them simultaneously. On the other side the two pieces of the wavefunction form a superposition which has a very different looking probability distribution (due to interference between its two pieces). By the time the wavefunction reaches the screen the distribution is such that the atom has a high probability of reaching certain places and no probability of reaching others. Although the atom will only manifest itself in one place, statistically many atoms, each with a similar probability distribution, will build up to form the pattern seen. It is important to note that this figure depicts nothing more than the evolution in time of a mathematical quantity, rather than the physical atom. Most physicists insist that it is quite wrong to think of both the wavefunction and the atom moving along as separate physical entities. Thus we are still faced with the problem of what the atom does on encountering the slits. Various ways of dealing with this issue will be discussed in Chapter 6.

– splits in two with each piece going through one of the slits. Note that I am describing here the way a mathematical entity is changing, and it is by solving the Schrödinger equation that I can tell what the wavefunction looks like at any given time. I can never know what is really going on, or even be sure there is anything real going on at all, since I would have to look to check, but as soon as I look I will alter the outcome.

As it goes through both slits, the atom's wavefunction is a superposition of two parts, each of which has the largest amplitude

at its corresponding slit. Now, if the state of the atom were described by just one of these parts of the wavefunction then we would say that it was definitely going through that slit. But in fact, the superposition of the two parts means that there is equal likelihood of it going through either slit.

On the other side of the slits, each piece of the wavefunction spreads out again and both sets of ripples overlap in such a way that their combined effect on reaching the screen gives the characteristic stripy pattern that we would see when two real waves interfere. Only now, we are not dealing with a real wave that washes up against the screen but a set of numbers that provide us with a probability for the arrival of a single particle at a given location.

Before the particle hits the screen, the wavefunction is all we have to describe reality. The wavefunction is not the atom itself but only our description of how it behaves when we are not looking at it. And it gives us all the information we can hope to obtain about the state of the atom if we were to look. So, provided we deal only with what the wavefunction looks like at any point, and follow the rules about how to use it to predict the chance of the atom being there and having certain properties, we can do business quite happily. Almost all physicists work in this way. This is because they have given up hope of ever being able to figure out what is actually going on using concepts rooted in Newtonian mechanics.

This is presumably not good enough for you. After all, it is all very well forgetting about the atom when we are not looking at it and dealing solely with the maths, but the fact is that a tiny localized particle leaves the atom gun, somehow ceases to behave like a particle for a while, and then pops up as a particle again at the screen. The route it must have taken is something we can only understand if we consider a wave going through both slits at once. All we can hope to do is track the atom's progress with the wavefunction. But the interference pattern that builds up after many atoms have been through the slits is real enough. Surely some sort of physical wave is going through the slits.

The problem here is that you want me to explain to you how the

atom goes through both slits by using images and concepts familiar to you from everyday experience. Unfortunately this is not possible. Whether we like it or not, such weird behaviour is a feature of the quantum world that we must accept, no matter how hard it is to believe. It does happen, and although we have a right to expect a rational explanation, none has yet been found. Many physicists will make statements such as: the world of atoms and below is so far removed from our own experiences in the macroscopic world that we have no right to expect things to behave in a way we can describe using everyday concepts. I know this does not sound very helpful, and even a bit of a cop-out. We should be disturbed by the way the atom behaves. But many of the greatest physicists believe this to be a dangerous and futile pursuit, and that the worrying is best left to philosophers, who have nothing better to do with their time!

Some physicists may criticise a non-technical book such as this, which over-emphasizes the mystery of quantum mechanics instead of stressing its accuracy and power in explaining so many phenomena. Well, that part will come later. But as for the two-slit trick, why can I not simply say that the atom somehow behaves like a spread-out wave as it passes through the two slits and be done with it? After all, if that is the way atoms behave then so be it. My response to those physicists who claim not to be disturbed by quantum mechanics is to insist that they have simply become desensitized to its implications through over-familiarity with the subject!

I will next briefly describe a device known as an interferometer, which highlights the idea of superposition (of a quantum system being in two states at once) in its purest and most baffling form. No longer can we even loosely think of an atom as behaving like a physical wave when we are not looking. Rather than send atoms through a screen with two slits, they can be sent one at a time through such a device.[5] On entering the interferometer, each atom is faced with a choice of two different paths, or arms, each of which follows a quite separate

[5] In practice, it is easier to use subatomic particles or photons of light (see box on page 76) but atoms will do for our discussion.

Particle interferometers

Interferometers are devices that highlight the way a single particle can travel along two paths at once and, once brought back together again, give rise to an interference pattern or some other type of signal that provides proof that something must have travelled along both routes.

Interferometers can be used with photons, electrons, or neutrons, and the way the wave corresponding to any of these particles is 'split up' once it has entered the device depends on what sort of particle it is.[4] For instance, photons are sent through optical beam splitters that behave like half-silvered mirrors. A photon impinging on a such a mirror placed at 45 degrees to its path will have a 50 per cent chance of passing straight through the mirror and continuing in its original direction, and a 50 per cent chance of being reflected into a path at right angles to the way it came in, just as it would if it encounter two slits in a screen, and so it takes both routes at once. On each side, further mirrors can guide the two paths away from each other, and later back together again.

Since the intensity of the beam can be turned down so that we can be sure that just one photon is ever in the interferometer at any given time, each individual photon will thus find itself in a superposition of travelling along the two independent paths at once, like the two paths in Robert Frost's poem. These two arms of the interferometer can be arbitrarily far apart – in practice they can be several metres apart. This makes it particularly difficult for

[4] Nowadays, physicists carry out experiments using atom and even molecule interferometers.

us to envisage the photon just as a wave. Its wavefunction really seems to now be in two isolated pieces.

Maybe we are forced to admit that the photon has either gone one way or the other and not both at once. Not so! The two arms of the interferometer are later brought together and recombined, and the signal we see emerging from the interferometer is a result of two waves interfering. Just as in the two-slit trick, the interference is due to the photon having to cover two different path lengths. The only way to explain what is seen is if the photon travels along both routes at once and ends up interfering with itself as it emerges!

Of course, if we place a detector along one of the arms to see if the photon goes along that route then we will see it half of the time, and the interference effect will disappear since now it has gone one way or the other.

Another way of saying this is that the interference pattern is produced when the two paths are indistinguishable. But if we insert a device in one of the arms that rotates the polarization of the photon wave in that arm by 90 degrees, then the two paths will produce distinguishable photons and the interference pattern disappears. This is because we will now have 'which-way' information about the photon once it emerges. The delayed choice aspect of the experiment refers to the fact that we need not even switch on the polarization rotator until the photon has split up into its two components (via a beam splitter or half-silvered mirror). If the two photon paths are vertically polarized initially, then the device will rotate the component in the upper arm to horizontal polarization. So if we detect a vertically polarized photon coming out we will know it must have travelled through the lower arm. If, on the other hand it is horizontally polarized then it must have been through the polarization rotator in the upper arm. Does this mean the photon only ever went one route or the other? And if it did, how could it have guessed that the rotator would be switched on after it had split in two inside the interferometer. After all, had we left the device switched off we'd have seen the interference pattern!

In 1982, physicists Marlan Scully and Kai Drühl proposed an even more astonishing extension to this idea. They suggested that even with a 'which-way' marker such as the polarization rotator in place and switched on, the information about the route taken by the photon can subsequently be erased just before the photon emerges. They proposed placing a 'quantum eraser' after the two paths have recombined (say via a second half-silvered mirror). You might quite sensibly think that once the two paths are distinguishable via their different polarization orientations, no interference would take place. But it seems that by removing all the evidence – by rotating the polarization by a further 45 degrees so that it is not possible to tell which route the photon took – the eraser can restore the interference pattern. This seems astonishing: not only does it appear as though the photon knows the rotator is switched on in one arm, but also that further down the line a quantum eraser removes the 'which-way' information.

An experiment was finally performed by Yoon-Ho Kim and collaborators a few years ago demonstrating Scully and Drühl's original scheme. The quantum eraser did indeed restore the interference pattern!

and independent route through the apparatus. Quantum mechanics tells us that, until we look, the atom's wavefunction will be in a superposition of two 'pieces' travelling along both routes at once. In principle the two paths can be very far apart, even on either side of the galaxy, but we must still regard the wavefunction as travelling along both routes. Finally, if we recombine the two paths we will see interference of a form that proves that the atom had to have gone both ways at once.

Interferometers show that quantum particles really can be in superpositions of two places at once. Of course, while I have not discussed it here, quantum particles can be in superpositions of other states, such as spinning in two directions at once or having two or more different energies or velocities at once. And while it might be comforting to say that it is really the wavefunction that is in a superposition and not the physical particle it is describing, something has to be travelling along both arms of the interferometer. Physicists often use contradictory or sloppy language to describe this situation by saying that there are two beams in the interferometer which interfere with each other. But what does this mean when we are describing just one particle? The truth is that no one can really explain it satisfactorily in non-mathematical language.

What we do know is that the atom is always described by a single wavefunction, and not two separate ones, that propagates through both arms. This is where we come unstuck if we insist on trying to think of wavefunctions like classical waves. If a sound wave is split up so that it travels along two different routes that are recombined at the end, we would also be able to observe an interference effect (by slightly altering the frequency of one of them we will be able to hear 'beats' due to the two waves ending up out of phase). However, the sound wave here really does physically split into two pieces. If the two arms carrying the sound waves end up in separate locations then two observers will each hear a sound. In the case of the atom, remember only one of them would ever see a particular atom if they looked for it. Formally, we say that the atom has just a single wavefunction with two pieces that describe how it travels through each arm no matter

how far apart they get. The wavefunction is spread out over the whole of space and has a value of zero everywhere apart from inside the two arms. The whole spread-out wavefunction then collapses into a single real particle in one arm or the other when we look.

Nonlocality

We are all familiar with the unsubstantiated, yet intriguing, claims of identical twins that are able to sense each other's emotional states even when separated by vast distances. Somehow, the idea goes, twins are joined at some psychic level that science has yet to understand. Similar claims would be made to explain how a dog senses when its owner is coming home, or how black magic voodoo dolls are supposed to work. I must stress that I do not mention these examples as phenomena that have any connection with quantum mechanics, nor do I believe they genuinely occur. I mention them only as wacky examples of a phenomenon called nonlocality. What is of interest here is that nonlocality has been shown beyond any doubt to exist in the quantum world, through an effect known as entanglement.

Consider a pair of dice. What is the probability of throwing a double? The maths to work this out is very straightforward. For any given number that one of the dice lands on, there is a one in six chance that the other die will land the same. Hence the chance of throwing two doubles in a row will be one in thirty-six (since 1/6 x 1/6 = 1/36). Of course this does not mean that if you throw a pair of dice thirty-six times then there will be exactly one occasion when two successive throws are doubles, it simply means that 'on average' this is the likelihood.[6] It follows therefore, by multiplying fractions, that the probability of throwing a double ten times in a row is about one in sixty million! This means that if every person in Britain were to have

[6] In the same way, tossing a coin twice does not guarantee getting both a heads and a tails just because they each have a 50-50 chance.

a go at throwing the dice ten times in succession then, statistically, just one person will get all doubles.

What if I were to give you a pair of dice that always land the same? Maybe first on a double six, then a double two, and so on – the actual number they land on is random but they are nevertheless synchronized. You would rightly be amazed. So you try to rationalize how the trick is being done. It could be that they are weighted by some clever internal machinery that controls how they land and that both dice have been pre-programmed to land on exactly the same sequence of numbers. This can easily be tested by throwing just one of the dice while holding the other. Now they are out of sync and the trick shouldn't work.

If they persist in landing the same despite this, then the only explanation can be that they must somehow re-synchronize before each throw by exchanging a remote signal. However, such signalling is subject to a crucial proviso: if the dice are very far apart (say one on Earth and the other on Pluto) they must both be thrown according to some prearranged timing such that there is not enough time for any signal to travel between them.

Of course, just throwing them once and then confirming later that they had landed the same could be down to luck. But repeating the process again and again on both Earth and Pluto, say once a minute, and still finding that they always land the same would imply some kind of instantaneous connection. Naturally, we could make sure there had been no prior signalling between the two to get them synchronized in the same way that we ensured they were not identically preprogrammed. We do this by throwing the Earth die an arbitrary number of times in the final minute leading up to the start of the experiment.

It takes light several hours to travel between Pluto and Earth and so it is impossible for the dice to communicate with each other via any known physical signal before each throw. If they persist in landing the same then we would have to concede that some sort of faster-than-light signalling is taking place, something not allowed by any known laws of physics.

In his Special Theory of Relativity Einstein showed how no object

or signal can travel faster than the speed of light. So you can imagine his disbelief when it was claimed that quantum particles really are able to communicate with each other in this way.[7]

The above discussion with the dice is an example of what is technically termed as nonlocal connection. By this I mean something happening over here instantaneously affecting something over there. This is what would be necessary if the two dice persisted in knowing what each other was going to land on too quickly for a signal to pass between them. In classical mechanics, this is not possible. The concept of cause and effect not only requires causes to always precede effects, but it is also subject to a strict condition. One of the most important lessons we have learnt from Einstein's theory of relativity is that if two events, one of which is the cause of the other, are separated by some distance then they must also be separated in time subject to the speed of light barrier. That is, if someone were to be hurt in a car accident, then the laws of classical physics state that his identical twin a thousand kilometres away cannot (no matter what psychic message might get sent) know about it faster than the time it takes for light to travel the distance between them – just a few thousands of a second.

Physicists are no longer in any doubt that instantaneous communication between distant objects, or nonlocality, is a general feature of the quantum world, and can be traced back to the nature of the wavefunction itself. Most physicists are not too bothered by this since quantum nonlocality can never be used for faster-than-light signalling – in violation of relativity theory – due to the inherent probabilistic nature of the quantum world.

We do not need the hypothetical example of the two magic dice to show nonlocality in action. It was a feature of the split-up wavefunction in the two arms of the interferometer we met earlier. If the two arms are light years apart, we can still decide to turn on a detector after the atom has entered the interferometer to check if

[7] This has nothing to do with Einstein's famous remark about God not playing dice. Remember that had to do with the possibility that the quantum world was ruled by chance and uncertainty. The fact that I have used a pair of dice in my discussion is coincidence.

it is in one of the arms.[8] If we detect the atom we would require the part of the wavefunction that is travelling through the other arm to instantly drop to zero as there is now no probability of the atom having gone that route.

The above property of the nonlocal collapse of parts of spread-out wavefunctions when we have a peek at what is going on somewhere else is known as just that: the collapse of the wavefunction.

I should point out that my example of the dice differs from what happens in quantum mechanics in an important way. If the two dice were entangled quantum-mechanically then throwing one of them alone, as I suggested to do to rule out any pre-synchronization, would always unavoidably alter the other.

Entanglement

So far in this chapter I have discussed two different and quite tricky concepts: superposition and nonlocality. The first states that a quantum particle can be in a combination of two or more states at the same time, while the second says that two quantum particles (or two separate parts of the spread-out wavefunction of the same particle) can somehow remain in touch with each other however far apart they are. I will now combine these two ideas together in order to introduce a third quantum concept.

In quantum mechanics, the idea of two dice remaining in (nonlocal) contact with each other how ever far apart they are is known as entanglement. While the term was first used in the early days of the field by Schrödinger, it is only recently that it has taken centre stage as a fundamental concept.

If two quantum particles interact with each other they can become correlated in the sense that their fates will be intertwined for ever,

[8] Of course in reality this would make for a very long experiment since we would have to wait several years for the atom to travel through our giant device so that the two parts of its wavefunction really are light years apart.

however far apart they get – until, that is, one of them interacts with a measuring device. The way this manifests itself mathematically is that they would be described by just one unified wavefunction that contains the combined and shared information about both their quantum states. Now it is possible for one of the particles to be put into a superposition, for instance if it encounters a double slit. If this happens, the second particle will also be forced into a superposition of different states that depend on (or, more technically, are correlated with) each of the two alternatives of the first particle. The wavefunction is now said to describe an 'entangled state'.

The most famous example of this was first described in a paper by Einstein and two colleagues, Boris Podolsky and Nathan Rosen. Of course Einstein was at the time more troubled by the concept of indeterminacy, whereby quantum mechanics insists that we cannot know everything about a quantum particle at the same time. However, we will see that it is the nonlocal character of entanglement that was the real issue.

The EPR experiment

In 1935, soon after he fled Germany to the United States, Einstein set about devising a thought (or *gedanken*) experiment to highlight quantum mechanics' insistence that we cannot consider a particle to have a well-defined position unless we look at it. Together with Podolsky and Rosen, he came up with a scenario that became known as the EPR Paradox, after the initials of the three authors of the paper. Put simply, they considered two quantum particles, such as photons, that are simultaneously produced by a common source, and which move apart with equal and opposite velocities.

If we focus on just one of the photons then we know by now that we must consider it as being like a spread-out wave until it is detected. We are forced to admit this because we know that if it encounters a screen with two slits then where it ends up is dictated by the interference pattern in its wavefunction. (Remember we can only see

the interference pattern when we have many identical photons going through the slits.) Each behaves as a wave before it is detected and a particle when it is detected. Einstein and his collaborators pointed out that it is the other photon that is of interest. If we choose to measure the wave-like properties of the first photon, such as its wavelength, then this amounts to measuring its momentum.[9] And since the two photons have the same magnitude of momentum – but travelling in opposite directions – then we will also know the second photon's precise momentum and can assign to it a precise wavelength. This amounts to saying that it is also behaving as a wave.

But, and here is the clever part, if we had instead chosen to measure the precise position of the first photon then it would appear to us as a localized particle. Now, we would also know the precise location of the second photon at that moment without looking at it, since it would have to have travelled an equal distance away on the other side of the source. Thus, we might be willing to concede that what attributes we assign to the first photon depend on what we choose to do to it and how we measure it. After all, we know we must be disturbing its wavefunction by measuring its position, as we saw in the two-slit trick. But we have not touched the second photon throughout. We could have presumably waited until the two were very far apart to be certain that we would not be disturbing the second one at all. The conclusion is that we could, in principle, have known either the undisturbed photon's precise position (particle nature) or its precise momentum (wave nature) at any given time without looking at it. It doesn't matter that we can't ever do this in practice, since we would have to have carried out the two different measurements on the first photon at the same time, what is important is that the second electron must have had a definite position and momentum all along.

I will return to this issue in Chapter 6 when I discuss Einstein's insistence on what is called an 'objective reality', namely that we

[9] Remember the discussion in Chapter 2 of de Broglie's formula relating the momentum of a particle to its wavelength.

shouldn't have to wait to measure a particular property of a quantum system for that property to be real. It is now known that he was wrong on this point (see box). Most physicists would say that neither particle even has a definite position or momentum until it is measured. While this view is not the only possible one, what is not in doubt is that there has to be some kind of nonlocal communication taking place, that instantaneously informs the undisturbed particle of the type of measurement we have just carried out on its partner.

Here is how the paradox is 'resolved': Since the particles have interacted, they will thereafter be described by an entangled wavefunction and their fates are linked however far apart they become. By measuring some property of one of them, the full wavefunction collapses and the second photon is instantly endowed with the corresponding property. Easy really, isn't it?

Today, quantum nonlocality and entanglement are no longer the subject of philosophical debate. They are accepted as crucial features of the quantum world. Indeed, entanglement of many particles could lead to the development of a whole new technology not even dreamed of by the quantum pioneers.

The quantum jewel thief

There are many examples of how quantum weirdness can be harnessed for novel applications but probably none more ingenious than the idea of interaction-free measurements. The following fictional scenario is a variation on a concept known as the Elitzur-Vaidman bomb-testing experiment, which has been verified in the lab.

An international jewel thief prepares for the ultimate heist. Locked inside a vault in a high-security Swiss bank are several priceless diamonds. Getting into the vault is by no means easy, but not an insurmountable challenge for our clever thief. Once inside, however, he is faced with what appears to be an impossible task. He has several hours inside the vault provided he does not try to remove a diamond. Once he starts to remove one though, he has just 30 seconds to complete the task and make it out through the vault door before it locks him

in. Since removing each diamond takes 20 seconds he can only ever steal one of them. The problem is that the security conscious bank has placed the diamonds among a hundred fake ones that look identical to the real thing. Since he can only remove one he must choose carefully. There is only one way to find out and that is by shining a special blue light on the diamonds: a fake will reflect the light whereas a real one absorbs it.

But even here the bank has been ingenious. If a real diamond absorbs just a single photon of the blue light it will be destroyed. It seems the owners would rather no one got their hands on the jewels than have them fall into the wrong hands. Of course only the bank knows where they placed the real diamonds and so have no need for the blue light test.

It would appear that there is no possible way the thief can succeed. Those that reflect the

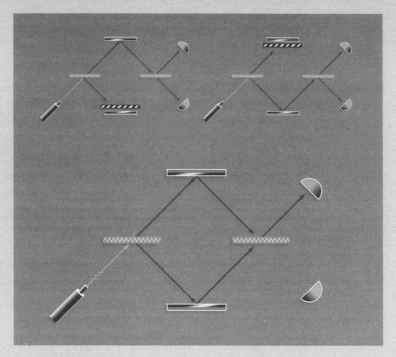

The Mach-Zehnder interferometer is composed of a photon source that fires one photon at a time, two half-silvered mirrors to split and combine the beam, two fully reflecting mirrors, and two photon detectors.
Top: If either of the two reflecting mirrors is blocked, there is only one surviving path (which the photon takes half the time) and the second half-silvered mirror ensures that there is an equal chance of the photon reaching either detector.
Bottom: With both mirrors able to reflect, the two paths can be adjusted so that interference between them in the second half-silvered mirror is such that the photon only ever reaches the upper detector.

light from his special blue torch purchased from the Jewel-Thieves-Accessories Store are duds and he can ignore them, but he cannot be sure without destroying it if he has chosen a real one. And he knows that the odds are only about one in ten

that he will make a lucky guess.

This is where a neat trick called a quantum non-demolition test can be used. All the jewel thief needs is a certain type of device known as a Mach-Zehnder Interferometer (see diagram). I will first outline the principles of this device.

When a photon enters the interferometer it encounters a half-silvered mirror that acts as a 'beam splitter' dividing the photon wave into two components, one that is transmitted through the mirror and the other reflected. Both parts bounce back off mirrors and recombine at a second half-silvered mirror. If the two intermediate mirrors are positioned carefully, the interference between the two paths can be tuned such that the photon always emerges in a certain direction. Of course placing a detector next to one of the middle (fully reflecting) mirrors so that we can see which way the photon goes, then as with the two-slit trick, this causes the quantum interference to be destroyed. Now each photon will not travel both routes at once but one or the other. In half the cases, the second half-silvered mirror ensures that the photon exits the device in the same direction it would have had no measurement been made. However, the other half of the time photons will exit in another direction, triggering an external detector to beep. So to recap, in the absence of measurement, the beeper never goes off, but if one of the mirrors is being watched the beeper will go off half the time since there will be no destructive interference stopping the photons from coming out in that direction.

The jewel thief's device simply uses the diamonds instead of the lower mirror. A fake diamond will reflect the light and the beeper never sounds. A real diamond, however, acts as a measuring device. Half the time the photon will travel along the path taking it to the diamond, and therefore destroying it. Never mind. The other half of the time the photon travels the other path. Here it is faced with a choice at the second half-silvered mirror. It is either reflected upward away from the beeper, or transmitted through and into the beeper. So, for every real diamond, the beeper will sound one quarter of the time. This signals that a real diamond is being tested, but one which is not destroyed because the photon was not measured along that path! Note the subtlety here. A fake diamond does not carry out a measurement since we do not gain 'which way' information from it.

As soon as the thief hears the beeper he simply has to

The lower mirror of the interferometer is now replaced with a diamond.

Top: A fake diamond reflects the light just as the original mirror would have done. The two paths interfere and the lower detector never registers.

Bottom: A real diamond will absorb the photon and be destroyed. But this constitutes a measurement as we now know which route the photon has taken, and half the time with a real diamond in place the photon will take the other path, avoiding the diamond. Since there is no interference now, the lower detector will sometimes register. When it does, we will have determined that we have an intact real diamond without the photon ever going near it!

remove that diamond and make his getaway. He has carried out a quantum measurement without ever touching the object he is measuring. I wonder if Hollywood would be interested in a movie based on this. I would be more than happy to play the quantum physicist-come-jewel thief – ably assisted by Brad Pitt and George Clooney of course.

The EPR Paradox and Bell's Theorem

In its original form, the Einstein-Podolsky-Rosen experiment was meant to show that quantum mechanics provided an incomplete description of reality; that all its weird aspects come about due to our lack of a complete understanding of the subatomic world. To place it in its historical context, the arguments presented in the EPR paper were part of a long-running debate during the 1920s and 1930s between Einstein and Bohr, the two giants of 20th-century physics. At the time of course, it was regarded as no more than a philosophical difference of opinion since no one knew how to perform such an experiment in practice.

Then in 1964, an Irish physicist by the name of John Bell proposed a way of testing who was right once and for all. Bell's theorem, or Bell's inequality as it is sometimes known, is an important milestone in the slow progress towards a full understanding of quantum mechanics. Indeed, it is regarded by many as one of the most profound scientific discoveries of the 20th century. Many popularized accounts of quantum mechanics go to great lengths to discuss it in simple non-mathematical terms. But the truth is that Bell's theorem cannot be described in just a few lines, nor without a good deal of careful thought on the part of the lay reader.

I will, however, outline the basic idea. Einstein had maintained that the reason two EPR-setup particles are measured to have correlated properties was hardly surprising. After all, they had been in contact in the past (as they were emitted by the same source). No faster-than-light signalling between them need be invoked if their properties had been set from the start. These well-defined properties, which quantum mechanics does not describe in the absence of

measurement, are known as 'hidden variables' and do not require any nonlocality, which Einstein was understandably nervous about. But was Einstein right? Could such hidden variables explain quantum weirdness?

Bell's theorem shifted the argument about the nature of quantum reality from the realm of philosophy to experimental physics.[10] He derived a formula showing that if Einstein was right then there would have to be a maximum amount of correlation between the two particles. That is, since the particles could have no prior knowledge of the type of measurement that would be performed on each of them, there was just so much conspiratorial preparation there could be built into them. Thus even with hidden variables that predefine both particles' properties, there is a limit to the synchronization between the results of measurements on the two particles. If, on the other hand, quantum mechanics and the concept of a single wavefunction describing the entangled state of the pair was correct then there would be a larger amount of correlation, or cooperation, than this maximum – in violation of Bell's inequality.

I will not go into the details of what sort of experimental set-up is required to test Bell's inequality, not because it is too technical but rather because analysing the data produced and drawing from it the astounding implications takes more than a few pages. Nevertheless it is quite straightforward and has been described clearly and in great detail elsewhere.

In 1982, a team of physicists in Paris, led by Alain Aspect, finally succeeded in carrying out the EPR experiment and testing Bell's theorem. They used two photons emitted from a calcium atom, which were correlated in the way they were polarized (at right angles to each other). Their results convinced most physicists that Bell's inequality is violated and hence that quantum mechanics, with all its weirdness, was in fact

[10] For a detailed, yet non-technical, discussion of Bell's theorem I recommend Dan Styer's *The Strange World of Quantum Mechanics* (Cambridge University Press, 2000). In fact, Bell also had a second theorem (known these days as the Bell-Kochen-Specker theorem) which is even harder to explain without formal mathematics. Brave readers can find out more about it in a review paper by David Mermin (*Reviews of Modern Physics*, Vol.65, 1993, page 803).

the way nature behaved. Had Einstein been alive he would no doubt have finally admitted defeat. Quantum mechanics really is nonlocal or, as Einstein preferred to say, it involves 'spooky action at a distance'.

Many physicists would of course prefer not to talk of such nonlocality. They will say that this is only necessary if we insist on looking for a physical mechanism to explain the experimental results. All we can say, they will insist, is that the measurement on each particle reveals some aspect of that particle's nature that was not known before. Not only can we not infer from the measurement what properties the particle had before it was measured, but it did not even have definite values for those properties; rather, it was sitting in a superposition of all possibilities waiting for a measurement to force it to make up its mind and consequently, through entanglement, the mind of its distant partner! However, despite such pragmatic reassurances, non-locality does not go away, it is just that many prefer to relegate it to the abstract mathematics by denying the existence of a physical instantaneous connection between the particles. This view is best put by Cornell Physicist N. David Mermin:

'Most physicists, I think it is fair to say, are not bothered [by the experimental confirmation of the violation of Bell's inequality]. A minority would maintain that this is because the majority simply refuse to think about the problem, but in view of the persistent failure of any new physics to emerge from the puzzle in the half century since Einstein, Podolsky and Rosen invented it, it is hard to fault their strategy. The mystery of the [EPR experiment] is that it presents us with a set of correlations for which there simply is no explanation. The majority would probably deny even this, maintaining that the quantum theory does offer an explanation. That explanation, however, is nothing more than a recipe for how to compute what the correlations are. This computational algorithm is so beautiful and so powerful that it can, in itself, acquire the persuasive character of a complete explanation.'

Quantum Chaology

Sir Michael Berry
Royal Society Research Professor, University of Bristol

The quantum world appears very different from the world of classical physics that it superseded. Quantum energy levels, wavefunctions and probabilities seem incompatible with Newtonian particles moving along well-defined orbits. Yet the two theories must be intimately related. Even the Moon can be regarded as a quantum particle, so there must be circumstances – roughly large, heavy objects – where the quantum and classical predictions agree. But the 'classical limit' is subtle, and much current research is aimed at understanding it.

Difficulties with the classical limit are extreme when the Newtonian orbits are chaotic. Chaos is instability that persists, so that motion, although strictly determined, is so sensitive that prediction is effectively impossible. With chaos, there is no regularity, no strict repetition. The weather is a familiar example. Another is the erratic rotation of one of the satellites of the planet Saturn, Hyperion, a potato-shaped rock about the size of New York City.

Chaos is problematic because the way a quantum wave develops in time is determined by the associated energy levels. A mathematical consequence of the existence of energy levels is that quantum time-development contains only periodic motions with definite frequencies – the opposite of chaos. Therefore there is no chaos in quantum mechanics, only regularity. How then, can there be chaos in the world?

There are two answers. One is that as the classical limit is approached – as objects get bigger and heavier – the time taken for chaos to be suppressed by quantum mechanics gets ever

longer, and would be infinite in the strict limit. However, this explanation fails because the 'chaos suppression time' is often surprisingly short – just a few decades even for Hyperion, a short time on the astronomical scale.

The true reason for the prevalence of chaos is that large quantum systems are hard to isolate from their surroundings. Even the 'patter of photons' from the Sun (whose re-emission gives the light by which we see Hyperion) destroys the delicate interference underlying the quantum regularity. This effect, of large quantum systems being dramatically sensitive to uncontrolled external influences, is called decoherence. In the classical limit, the quantum suppression of chaos is itself suppressed by decoherence, allowing chaos to re-emerge as a familiar feature of the large scale world.

Smaller quantum systems, such as atoms in strong magnetic fields, molecules vibrating strongly, or electrons confined in 'quantum dots' with unsymmetrical boundaries, can be effectively isolated. Therefore decoherence is irrelevant and there is no quantum chaos, even though the corresponding classical systems are chaotic. Nevertheless, these quantum systems reflect classical chaos in several ways, whose systematic study is quantum chaology.

The energy levels of highly excited states form a set of numbers that can be studied statistically. These statistics (for example, the probability determining the spacings between neighbouring levels) are different when there is chaos and regularity. Similarly, the patterns of the quantum waves describing the states are different. A surprising and still-mysterious discovery is that the arrangement of energy levels in quantum chaology is related to one of the deepest problems in mathematics, involving patterns in the prime numbers.

Chapter 5
The Watchers
and the Watched

The basics of quantum mechanics I have outlined so far might have sounded baffling to you, even far-fetched at times, but the fact is that, mathematically and logically, the quantum rules are unambiguous and well-defined. And while many quantum physicists themselves do not feel comfortable about translating the strange abstract properties of the wavefunction to the real world, the mathematical framework and formalism of quantum mechanics has been too successful and is too accurate for there to be much doubt that it is reflecting fundamental truths. But there remains one final puzzle that quantum physicists have been unable to explain satisfactorily. Many would argue that it is the most important and yet most puzzling aspect of all, namely: why is it we can never watch the wavefunction – or whatever physical reality it describes – in action when we look? Or, to put it another way, why does the interference pattern disappear when we try to probe which slit the atom goes through? These questions have not been answered by quantum mechanics, and form what is known as the quantum measurement problem.

We see the effects and influence of the wavefunction with its probabilistic nature, nonlocality, and ability to form superpositions and entangled states all around us. In fact, we need these very properties to explain the solidity of all matter, how the Sun shines,

even how the atoms that make up our bodies were formed. But the basic problem remains: how do we understand the way a spread-out wavefunction suddenly transforms into a localized particle as soon as we try to observe it?

Quantum physicists refer to this mysterious process as the 'collapse of the wavefunction' – indeed, I used the phrase myself in the last chapter. A relatively recent discovery, however, has convinced many physicists that such terminology is unnecessary, but nobody is really sure yet whether the measurement problem has been solved or not.

I will begin by first discussing what we mean by 'looking' at something.

What you see is what you get

In our everyday world of macroscopic objects, I take for granted the notion that the way an object appears to me is exactly as it really 'is'. I am naturally assuming I can trust my eyesight, am not on hallucinogenic drugs, and that the lighting is sufficient. Indeed, to see something, light must either be emitted by the object or, more likely, reflected off it and captured in my eyes. The image formed on my retinas is then interpreted by my brain.

But has it occurred to you that by the very act of bouncing light off something we are disturbing it and thereby altering it in some small way, either by heating it up a tiny fraction of a degree, or by causing it to recoil ever so slightly from its original position? Naturally, when I look at a table or a car, or even a living cell under a microscope, the collision of the photons of light are not going to have any sort of effect I could ever measure. But when we deal with quantum objects, that are themselves on the same scale as the photons, things are different. Think back to your school physics and Newton's third law: every action has an equal and opposite reaction. In order to 'see' an electron we would have to bounce a photon off it. But by the time we capture the photon, the electron will no longer be where it would have been had it not been hit.

This unavoidable disturbance of a quantum particle when it is observed is often used to explain the problem of quantum measurement and even as the basis of Heisenberg's uncertainty principle. Not only is this too simplistic but it is also quite wrong and conjures up an image of classical balls bouncing off each other. While this picture was crucial in proving the particle nature of light, as we saw in Chapter 2 when I discussed the photoelectric effect and Compton's scattering experiments involving X-ray photons and electrons, it tells us nothing about the true quantum nature of either photons or electrons.

Nevertheless, disturbing an object by measuring some property of it is easy to understand. Here is another simple example. I determine the temperature of water in a bath by using a thermometer. Initially, there will be a small transfer of heat to the thermometer until it reaches the same temperature as the water, but this tiny heat loss by the water is not likely to affect its temperature (after all, there will be far more heat lost to the surrounding air than is used in warming up the thermometer). But measuring the temperature of water in a small test tube by immersing a thermometer will result in a relatively large amount of exchanged heat unless the thermometer is already at the same temperature as the water. The reading will therefore not be an accurate measure of the water temperature before the thermometer was immersed.

Thus to learn something about a system we must measure it, but in so doing we often unavoidably change it, and will therefore not know its true nature. This problem can usually be circumvented in the macroscopic world but not in the quantum domain.

Heisenberg's gamma-ray microscope

In Compton scattering, X-ray light is shone on a solid target – a sheet of graphite was used in the original experiment – and the reflected X-rays are analysed. It is found that the X-rays' frequency drops slightly after reflection. This was successfully interpreted by Arthur Compton

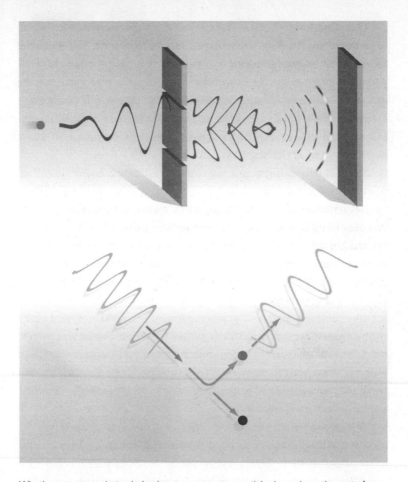

Whether we see a photon behaving as a wave or a particle depends on the sort of experiment we conduct. In the two-slit experiment (top), the photon starts off as a localized particle, behaves as a wave when going through both slits and interfering on the other side, and is finally detected as a localized particle again. In Compton scattering (bottom) it starts off as a wave with a definite wavelength, behaves as a particle bumping into an electron, and is finally detected as a wave again with a slightly longer wavelength due to the momentum lost in the collision.

(using Planck's relation between frequency and energy) as particles bouncing off each other; the electrons in the target are knocked out carrying away some of the energy of the incident X-ray photons.

This situation is the opposite of the one in the two-slit experiment with the atoms. In that case, an atom comes in as a particle, behaves like a wave in going through the slits, and finally ends up as a particle on the screen. In Compton scattering, the photon comes in as a wave (with some frequency), behaves like a particle in hitting the electron, and is finally detected as a wave again when its frequency is measured. In both experiments we make use of the notion of wave-particle duality of the atoms and the light.

But simply talking about wave-particle duality doesn't help us understand how the processes take place. Indeed, the phrase is a throwback to the early days of quantum theory and it is a shame that it still appears in the teaching of the subject today.

The reason I have described Compton scattering again is that it bears a striking resemblance to a thought experiment devised by Werner Heisenberg in the mid-1920s. With it, he was able to derive his famous uncertainty principle formula by highlighting how the act of observing a quantum particle disturbs it from its original path. Unfortunately for someone of the genius of Heisenberg he was off the mark here. Niels Bohr told him so in no uncertain terms at the time, even reducing him to tears on one occasion – these guys took their work very seriously. But to this day, Heisenberg's example remains both confusing and unhelpful.

The idea involved what he a called a gamma-ray microscope. To view something under a normal microscope we shine visible light on it, which is then reflected into the microscope's lenses. But this would be of no use to view an object smaller than the wavelength of the light itself (a few ten-thousandths of a millimetre) since such an object would be incapable of reflecting the light. But X-rays and gamma rays are forms of electromagnetic radiation with much shorter wavelengths and can hence be used to 'look at' smaller length scales.

Heisenberg's gamma-ray microscope was a hypothetical device that he suggested could be used to 'see' an electron using the Compton scattering idea. He argued, correctly, that in order to pinpoint the position of the electron, the gamma photon would have to strike

it and bounce back through the microscope lens. But in doing so it would impart to the electron a 'kick' that would alter the electron's momentum. By considering things like the resolving power of the microscope and the wavelength of the photon, he was able to derive his uncertainty relation. This states that the product of two numbers, one giving the uncertainty in the electron's position and the other the uncertainty in its momentum, will always give an answer that is greater than the value of Planck's constant. This means that, while Planck's constant is an incredibly tiny number, it is, crucially, not zero. We therefore always end up with some uncertainty, either in a particle's position or its momentum (or both). Either one of these quantities on its own can be measured exactly, but only at the expense of our knowledge of the other.

The problem with gamma-ray photons is that their very short wavelengths corresponds to them having, using the de Broglie formula, very high momentum. So the more precisely we pin-point the electron's position the greater the momentum kick we unavoidably give it. If we try to 'see' the electron with less energetic, and hence more gentle, light then we necessarily require the light to be of longer wavelength. But now we can no longer establish the electron's location so accurately.

We have seen, however, that the uncertainty relation was a consequence of the relationship between the position and momentum wavefunctions of the particle. While Heisenberg's example might seem far more simple and intuitive than talk of wavefunctions, he neglected a crucial point. His derivation relied on the wave-particle duality of the photons, while treating the electron all the while as though it were a point particle! Both the electron and the photon really need to be treated on an equal footing.

So where does this leave us? I can put it this way: we will unavoidably disturb something like an electron when we try to look at it, but this is not the origin of the uncertainty principle; it is an addition to it. The uncertainty principle is a far more fundamental feature of the quantum world and can only be understood in as much as we understand the nature of the wavefunction. Thus, the

uncertainty principle is not a result of our clumsiness in trying to pin down the electron when we try to look at it. After all, I am sure Isaac Newton himself would have had no problem admitting that we will disturb the state of something as tiny as an electron when we bounce particles of light off it.[1]

Well, now that I have cleared up the misconception that the only problem with the act of measurement is one that can be explained away using classical mechanics, we can think a little more about the real problem. It turns out to be much more fundamental than the concept of wave-particle duality and does not require us to resort to the uncertainty principle.

'And then something else happens'

When I discussed the concept of fate and predicting the future in Chapter 3, I ended by saying that quantum mechanics rescues us from the depressing Newtonian idea of a deterministic, clockwork universe where everything that will ever happen is mapped out and can, in principle, be predetermined. Instead of being able to predict the future with certainty by solving Newton's equations of motion we are left with only ever being able to predict the probabilities of different outcomes.

This is the essence of quantum indeterminism. But is Schrödinger's equation, which we use in the quantum world rather than Newton's equations, really the origin of this indeterminism? The answer is, perhaps somewhat surprisingly, that it is not. Indeed, Schrödinger's equation is perfectly deterministic: given the wavefunction at any moment in time I can compute it exactly by solving the equation for any future time. The notion of probability only comes in when we wish to put down our pen and paper, switch off the computer, and make predictions about the outcomes of a

[1] Remember Newton already believed that light was composed of particles.

real measurement using our knowledge of the wavefunction at the time of measurement.

For all its success and mathematical power, the formalism of quantum mechanics tells us nothing about how to take this step from the Schrödinger equation to what we see when we make a specific measurement of the quantum system being studied. For this reason, the founding fathers of quantum mechanics came up with a set of 'postulates': rules that are an added extra to the quantum formalism, which provide a recipe for how to translate the wavefunction to definite answers, or 'observables' – tangible properties we can observe, such as the position or momentum or energy of an electron at a given moment.

One such postulate you have already met states that the probability of finding a particle in a certain location is obtained by adding together the squares of the two numbers that define the value of the wavefunction at that position. This rule is not forced upon us by the mathematics, but it works. The other postulates are also to do with what kinds of measurements can be made and what we should expect to find when we carry out a particular type of measurement. They give us an ad hoc set of instructions on how to proceed when we have to abandon the 'deterministic' Schrödinger equation and compare its predictions with observations.

The concept of measurement in quantum mechanics is rather nebulous and ill-defined. So physicists have been happy to adopt the pragmatic view laid down by Bohr, who considered a measurement to involve a mysterious process referred to as the 'irreversible act of amplification' that takes place when a quantum system interacts with a measuring device. Crucially, the measuring device here must be described by classical physics and is therefore not a quantum object. But how and why and when does this measurement process take place? Any measuring device, whether it be the screen in the two-slit experiment, a voltmeter, a Geiger counter,[2] an elaborate

[2] This is a type of particle detector that makes a clicking sound every time it captures a subatomic particle. Geiger counters are used to measure levels of radioactivity.

machine with lots of knobs and dials, or even a human experimenter, is ultimately composed of atoms too. So where are we supposed to draw the line between what is a quantum system obeying the quantum rules and what is a classical system that qualifies as a measuring device?

Schrödinger's cat

In the same year that Einstein, Podolsky and Rosen published their famous paper highlighting what they believed to be the absurdity of certain features of quantum mechanics (indeterminism and nonlocality), Erwin Schrödinger had also had enough of quantum weirdness. He tackled the measurement problem by proposing one of the most famous thought experiments in the whole of physics. It is incredible how many physicists have tied themselves up in knots trying to emphasize or explain away the supposed paradox laid out by Schrödinger. Over the years, many have come up with fantastically exotic resolutions – as if quantum mechanics wasn't tough enough to cope with already.

Schrödinger asked what would happen if we were to shut a cat in a box with a device containing a lethal poison and a radioactive atomic nucleus. The particle emitted by the nucleus when it decays triggers a mechanism that releases the poison into the box, and the cat is instantly killed. (Animal rights activists are assured that no such experiment has actually been carried out since the same aims could be, and indeed have been, achieved without a live pussy cat.)

The moment of decay of a radioactive nucleus is another one of those events that cannot, even in principle, be predicted exactly. All we can say is that there will be a certain probability that a given nucleus will have decayed after a certain time. What we can say, however, is that at the moment we close the lid on the healthy cat we know that the nucleus has definitely not decayed. Thereafter, it has to be described by a quantum superposition. That is, its wavefunction

must be composed of two parts, describing both a decayed and undecayed nucleus. Of course, the probability of finding the nucleus in one or other state will change over time. After a very long time the part of the wavefunction describing an undecayed nucleus will make up only a tiny component, and the probability of finding the nucleus to have decayed will be almost equal to one hundred per cent, just as we might expect.

So far so good. But here, Schrödinger followed the letter of the (quantum) law. Since the cat is also composed of atoms it should also be described by a wavefunction – a very complex one of course, but a wavefunction nonetheless. And since the fate of the cat is now strongly correlated with that of the radioactive nucleus, we must describe the two by an entangled state. Therefore the cat's wavefunction will also unavoidably split into a superposition of two states: one describing a live cat, and the other a dead cat!

Let me again remind you of the two-slit trick. Unless we look to see where the atom is it does not go though one slit or the other, but must be regarded as having gone through both at once. This is not just mathematical trickery, but is the only way to explain the very real interference pattern we see. We have the same problem with Schrödinger's cat. Until we open the box to check on the state of the cat we cannot say that it is either dead or alive (that is equivalent to saying that the atom goes through one slit or the other) but must be in both states at once! Surely this is ridiculous. The cat must be dead or alive, and our opening of the box cannot influence the outcome. Isn't it just a matter of our ignorance of what has already happened (or not happened yet)? Well, this was precisely Schrödinger's point.

And if you feel like siding with Schrödinger on this then I should ask you where we are going wrong? Why isn't the cat – or the 'complicated collection of quantum particles' as I might wish to refer to it – also forced into a superposition through its entanglement with the superposed state of the radioactive nucleus?

Bohr and Heisenberg did not claim that the cat was really both dead and alive at the same time. They insisted instead that – and

this has been accepted by the majority of physicists ever since – we cannot talk about the cat as even having an independent reality until we open the box to check up on it! Their reasoning was that all the time the box is closed we simply have nothing to say about the 'real' state of the cat. All we have to go by is the wavefunction, and this is just a set of numbers. Since we cannot describe reality in the absence of measurement then we don't even try. Only when we are ready to open the box can we use the wavefunction to predict the likelihood of finding the cat dead or alive. And repeating such an experiment many times over would show that such predictions are correct (just as we would need to flip a coin many times to confirm the 50/50 statistical probability of heads versus tails).

But what if we were to place a human volunteer in the box instead – ok, maybe the poison is not fatal, but simply causes the volunteer to lose consciousness.[3] Surely, once we have opened the box we cannot then convince the volunteer that he was in a state of being both conscious and unconscious at the same time before we let him out. If he is conscious he would report that apart from feeling a little nervous he has been fine all along. And if we find him unconscious then he might, when revived, report how he heard the device go off ten minutes after the box was closed and immediately started to feel dizzy. The next thing he knew he was being revived with the smelling salts.

Where is the flaw in this argument? Was Schrödinger successful in highlighting an inconsistency in the theory? Actually, quantum mechanics poses a more serious dilemma than the simple statement that, prior to opening the box, the cat's wavefunction has a live cat part and a dead cat part. Until we look inside the box all we can do is assign probabilities to different outcomes. And because of the superposition of the different cat states, we get a certain probability for finding the cat alive, another probability for finding it dead and, strangest of all, a probability for finding it both dead and alive at

[3] Yes, I know I could have used the same reduced dose for the cat, but I was describing Schrödinger's original idea.

the same time. This is due to the technical definition of quantum probability.[4] But the quantum rules do not allow for this. They state, quite correctly thankfully, that we will only find a live cat or a dead cat and never a cat in a ghostly in-between state.

I don't know about you, but I can cope with only having probabilities for dead and alive outcomes; I can put this down to my ignorance and I can pretend it is no more serious than the situation of not knowing what Christmas presents I have until I unwrap them. I might even be prepared to relegate the issue of cats being both dead and alive at the same time before I resort to pointless metaphysical pondering. But what happens to that third option, the possibility of finding the cat in the somewhat embarrassing state of being dead and alive simultaneously, which the quantum formalism is so clear about? The quantum postulates get round the dilemma by stating that the only allowed results of measurements are 'sensible' ones. Of course this is all couched in appropriate jargon, but the fact remains that, until recently, nobody really understood what happens to these in-between 'Schrödinger cat' states (as they have become known).

Many, even modern, discussions of quantum measurement do not appreciate that the paradox of Schrödinger's cat may have finally been settled. They talk about the act of opening the box as being the moment of 'measurement' that collapses the superposition and leads to a definite outcome. At one time it was even fashionable to talk about a measurement requiring an intelligent, conscious observer. After all, since no one knew where to place the boundary between the quantum domain of wavefunctions and superpositions and the classical domain of definite outcomes when measurements are made, then maybe we should only place a boundary when we really have to. Since even the measuring apparatus (detector, screen, cat) is just a collection of atoms and should behave like any other quantum system, albeit a very big one, then the only place we

[4] Since the wavefunction is made up of two pieces then the square of their sum is not the same as the sum of their squares but is always larger by an amount known as the interference term. To see this, note that, for instance, $(2+3)^2=25$ but $2^2+3^2=13$.

are forced to abandon the quantum description is when it registers in our conscious minds. (And yes I know the cat is also conscious but supporters of this view argued that it didn't have a PhD and therefore didn't qualify as an observer!) There still exists a relatively widely held view, mostly outside of physics, that since we do not fully understand quantum mechanics and we do not understand the origin of consciousness then the two must be related.

I have always been surprised at the number of prominent supporters this view has had. One of the basic tenets of quantum mechanics is the fact that the results of measurements are defined not only by the properties of the quantum system being studied but by the very act of measurement itself. Many also go as far as to state that it is meaningless to say anything about the quantum system without including the measuring device in the discussion as well. But to place the dividing line between measured and measurer at the level of human consciousness amounts to what philosophers refer to as solipsism – the idea that the observer is at the centre of the universe and that everything else is just a figment of his or her imagination.

In this way, what the volunteer inside the box reports he felt before the lid was opened is not important. All that matters is that to you, outside the box, he remains in a quantum superposition until you open the box yourself, whereby your consciousness forces the wavefunction to collapse to one or other alternative.

Don't tell me the score

While chatting to my brother recently about quantum mechanics and football (what else is there?) he relayed what must be a familiar feeling many of us have that is quite similar to the role of the observer in quantum mechanics. You see, we are both lifelong supporters of Leeds United,[5] and he had often wondered whether, if he recorded

[5] I should explain to American readers that Leeds United is the top English soccer club – in my view of course.

This simple optical illusion shows in a rough way how the act of measurement affects a quantum system. If you stare at any one of the white circles you will notice black dots appearing and disappearing in some of the surrounding circles. But if you switch your gaze to one of these black dots it immediately turns white again and stays white. You can never catch a black dot in the act. Naturally, the way this trick works has nothing to do with quantum mechanics, but I think the analogy is quite neat.

a Leeds match on video and then sat down to watch the recording without knowing the score then, for him, the match does not yet even have an outcome. But if we push the standard interpretation of quantum mechanics to its logical conclusion then it seems to be saying that this is not just a matter of his ignorance of the final score; information that is 'out there' and known to millions of people. It is only after he has seen the end of the match that the superposition of all possible results collapses, for him, to the one outcome he measures: the result he sees.

However, until he calls me to tell me the score, assuming I do not know it yet, then my brother will be, to me, in a superposition of having knowledge of all possible scores. In hearing the news from him, I have in effect carried out a measurement on his quantum state and forced the superposition of the various happy and disappointed versions of my brother to collapse to just one.

Julie's words again ring in my ear as I write this. But I would have to agree with her that this really is a 'load of nonsense'! I prefer to think that there is an 'objective reality' out there that exists whether or not I look. To give you a more solid example: a radioactive uranium nucleus buried in the earth will emit an alpha particle that can leave a visible track of crystal defects in the rock. It does not matter whether we look at the rock today, in a hundred years time or never. The track will be there. What if the rock is on Mars and is never discovered by a conscious observer? Does it sit in a state of limbo, both having and not having a track etched into its structure? Clearly measurements must somehow be taking place all the time, and conscious observers, whether they wear white lab coats or not, cannot play a role. The correct definition is that a measurement is deemed to have taken place when an 'event' or 'phenomenon' is recorded. This would be something we could perceive, if we wish to do so, at a later time.

The above statement might sound so sensible and obvious that you could be forgiven for wondering how any quantum physicists could be stupid enough to hold an opposing view. But then again, if we have learnt anything about quantum mechanics it is that searching for rational explanations is a futile exercise. Many deep thinkers have struggled with the issues surrounding quantum measurement, and most of the differing views they end up holding cannot be refuted so easily.

Over the years, I have had many discussions with a theoretical physicist colleague of mine, Ray Mackintosh, with whom I do not always agree, but almost always concede that the points he makes are valid. Our most entertaining 'debates' tend to be in bars during evenings while we are away at nuclear physics conferences together. (This is the usual setting for most such deep discussions among theoretical physicists who do not have the 'pleasure' of the experimentalists' night shifts in the lab.) Mackintosh raises the following objection when he hears criticism of the view that nothing exists until it is measured (and its wavefunction collapses). Of course the electron, or atom, or Moon, exists prior to measurement. All objects, whether microscopic or macroscopic, have many well-

defined properties (i.e. quantities that are not in superpositions of more than one value), such as their mass or electric charge. These things are not subject to any uncertainty. So just because we must rely on the wavefunction prior to measurement does not imply that object it describes is not real. Of course, certain of its properties, such as exactly where it is or what energy it has, cannot be said to have definite values, but then that is just the nature of quantum objects. We should not dismiss their existence just because we cannot have a mental picture of what they are really like.

Two stages of the measurement problem

So what constitutes a measurement? What do we mean by an 'observation'? And where is that magical transition from wavefunction to actuality? We do know this much: a measurement must first involve an interaction of some kind between a measuring device, which I have not yet specified, and a quantum system. The latter could have until that moment been in a superposition of different states corresponding to different possible outcomes of the quantity being measured. This interaction will lead inevitably to an entangled state of a system plus external measuring device, forcing the latter into a superposition of having measured all the different possible values of the quantity at once. Then what?

Let us analyse a little more carefully the particle interferometer set-up of the last chapter. Remember that the wavefunction of a single atom will, on entering the device, split into two components, each travelling along one of the arms. And whilst I must describe the atom by a wavefunction, I know something strange must be happening since I see a real interference signal on the other side. I also know that if I look through a porthole in one of the interferometer's arms to see whether an atom is travelling along that route then I will detect half of the atoms. But now of course the interference pattern disappears.

To simplify matters I will assume that the way I detect an atom

is by opening a small porthole in the arm and seeing a tiny flash as light bounces off the passing atom. So what happens if I now decide to close my eyes? I now have no knowledge of which way the atom has gone; does this mean the interference pattern comes back again? This would imply that the physical act of moving my eyelids is controlling the outcome of the experiment. Not a very convincing argument. What now becomes obvious is that the measurement must somehow be taking place because I have opened the porthole and allowed light to come in and interact with the atoms. The interference disappears whether or not my eyes are open. Of course I could argue that the interference signal at the end is still a possibility (one of three actually: either the atom was detected in the arm, or it wasn't and must have gone through the other arm, or it was both detected – by the light coming in through the porthole – and not detected, at the same time). The third of these options is the Schrödinger cat state (the dead-and-alive-at-the-same-time option, that is) and leads to the interference signal in the interferometer. Of course I never see this option when I open my eyes, and the conscious observer supporters would still argue that the act of my seeing, or not seeing, the atom is what ultimately eliminates the option that gives rise to the interference.

The problem is made clearer if we make sure that the flash of light due to an atom going past the porthole leaves a permanent record in an instrument placed outside the interferometer. This now plays the same role as Schrödinger's cat in the box. Is the instrument capable of existing in a superposition of recording and not recording a flash?

All in all, we can identify two distinct issues relating to the measurement problem:

(i) Why is it that we never see superpositions of such macroscopically distinguishable states (such as cats being dead and alive at the same time, and instruments both recording and not recording signals)? After all, we do see the effects of such superposition in the interferometer when the porthole is closed, and of course in the two-slit experiment. But there the states involved were microscopic ones involving individual atoms.

(ii) Even if there was a way of getting these Schrödinger cat states out of the way before we look, is there not still further wavefunction collapse necessary to remove all the remaining possible options bar the one we actually end up observing? Thus, Schrödinger's cat could be dead, alive or both. Removing the 'both' option still does not tell us how one of the other two possibilities is discarded when we open the box.

In an atom interferometer device, the atom's wavefunction is forced into a superposition of travelling along both paths at once. But if we open a porthole to check which way it has gone, we force it to make up its mind. If we observe the atom then we instantly collapse the component of the wavefunction in the other arm to zero. Even if we don't see it, we still force it out of the quantum realm to become a definite particle, travelling now via the other arm.

Decoherence

During the 1980s and 1990s, the first part of the problem was cleared up. Physicists came to appreciate what must be happening when an originally isolated quantum system such as a single atom, existing happily in its superposition, becomes entangled with a macroscopic object. It turns out that the superposition of different states forced upon such a complex system involving a trillion trillion atoms simply cannot be maintained and very quickly disappears, or decoheres. One way to think about it is to say that the delicate superposition gets irretrievably lost among the stupendously large number of other possible superpositions due to the different possible combinations of interactions between all the atoms in the macroscopic system. Recovering the original superposition is a little like trying to un-shuffle a pack of cards so that all four suits are separated, only incredibly more difficult.

Decoherence is a real physical process that is going on everywhere all the time. It takes place whenever a quantum system is no longer isolated from its surrounding macroscopic environment and its wavefunction becomes entangled with the complicated state of this environment. This might be anything from a photosensitive screen or electronic device to the surrounding air molecules. If the coupling to this external 'environment' is strong enough then the original delicate superposition very quickly gets lost. In fact, decoherence is one of the fastest and most efficient processes in the whole of physics. And it is this remarkable efficiency that is responsible for decoherence having evaded discovery for so long. It is only now that physicists are learning how to control and study it. I will describe in Chapter 10 a few of the remarkable experiments that have been carried out over the past few years in which we can actually see decoherence in action.

Technically, physicists talk about the environment wavefunction losing all remnants of the initial phase correlations between its two entangled pieces. A more colourful phrase is simply to say that once

a quantum superposition gets entangled with the outside world, all the weirdness leaks out and gets lost so quickly that we never see it in action.

The process of decoherence is still an area of active research and is not yet fully understood. But we can at least begin to make sense of the first part of the measurement problem. The reason we never see Schrödinger's cat both dead and alive at the same time is because decoherence takes place within the box long before we open it. This is not even down to the cat but would take place much earlier, due to the device housing the radioactive nucleus and the poison. This is what forms the macroscopic environment immediately surrounding the radioactive nucleus.

Another example in which we can apply the idea of decoherence is the case where a detector registers whether an atom is in one arm of an interferometer or going through just one of two slits. Here, the explanation is easier. The detector must, in order to gain information about the atom's position, couple to the atom's wavefunction and the entanglement quickly leads to decoherence. Note that I said 'the atom's wavefunction' and not 'the atom' since the detector not registering would imply that the atom has gone via the other route. This is also a 'measurement' that will destroy the interference pattern. So, even though the detector and the atom need never physically come into contact in a classical sense, there is still entanglement with the wavefunction.

Without a detector in place behind one of the two slits, it is the screen at the back that causes the decoherence. Only now it is too late to stop the interference pattern from developing; the two parts of the atom's wavefunction have been left alone for long enough to interfere with each other. Of course the whole two-slit trick would have to be carried out in a vacuum, otherwise the atom would bounce around off the air molecules and betray its position. This amounts to saying that the air itself would bring about decoherence. But again I have been a little sloppy with my wording. By using phrases such as 'bouncing around' I have made it sound as though the atom were already a classical particle with a predefined position. We now know

that prior to decoherence and measurement we can only describe the atom in terms of its wavefunction.

Does decoherence solve the measurement problem?

Some would say no, it does not help us solve the second part, namely how, upon measurement, one outcome from a number of alternatives – all of which are possible results – is plucked out. Decoherence tells us why we never see a cat that is both alive and dead at the same time, but it doesn't tell us how one or the other is selected. Quantum mechanics remains probabilistic and this unpredictability of individual measurements does not go away.

Other physicists do not feel that decoherence adds much to the measurement debate at all, even at the level of getting rid of the superpositions. They argue that all that happens is that superpositions become buried in the complexity of the entangled wavefunction of quantum system plus macroscopic measuring device but are, in principle, still there. We are simply unable to recover them. And this is why we never see such quantum weirdness in action.

Those who feel more optimistic about the elucidation of the measurement problem brought about by an understanding of decoherence would see things differently. They would argue that while the probabilistic nature of quantum mechanics means we cannot, even in principle, say with certainty whether the cat is going to be found dead or alive, this is now equivalent to familiar statistical probability. It is simply our ignorance of which option has been selected since all we have at our disposal is the wavefunction. But the cat is already either dead or alive!

So, the coupling with the environment surrounding a quantum system that inevitably takes place when the system can no longer be considered as isolated, leads to entanglement and hence decoherence. Does this mean a measurement has taken place? If, in addition to the decoherence, one of the remaining permissible outcomes is picked out then yes, that's that; a measurement has taken place. There is

The way the process of decoherence causes quantum weirdness to die away can be shown by a 3D plot known as a Wigner distribution. The two horizontal axes denote position and momentum, and so each point on the bumpy surface corresponds to a particular position and momentum of a quantum particle. The two broad bumps denote the two possible locations of the particle if we were to look for it. Clearly it is twice as likely to be found on the right as it is on the left since that bump is twice as high. The oscillations in the middle correspond to the interference terms. These do not correspond to possible results of measurement – they are the cat-is-dead-and-alive-at-the-same-time states.

The upper diagram shows these interference bits to be significant since decoherence has yet to set in. But after a very short time (lower diagram) decoherence causes them to die away. But the two physically realizable options remain unaffected.

no need for an observer to collapse the wavefunction and a single outcome is there, waiting for us to verify it when we wish.

My feeling is that there is still more to be understood. I am optimistic that even the post-decoherence selection process will one day be explained. Some physicists, such as Roger Penrose, think that it can. Many others, working in the fields of cosmology and quantum gravity, think that the problem has already been solved.

So, next stop: the various interpretations of quantum mechanics.

Chapter 6
The Great
Debate

Formalism versus interpretation

An appreciation of many of the quantum concepts, such as the wavefunction and its strange properties, as well as the postulates that tell us how to extract information from it about the subatomic world, are essential for the success and understanding of the theory. And yet we have seen in the first half of this book how difficult it is to translate what is essentially advanced mathematics into words that make sense, to both physicists and non-physicists. Another way of putting it is that while the formalism of quantum mechanics is not in doubt, nobody has yet found a satisfactory explanation, or interpretation, of the theory that is agreeable to everyone.

For the two-slit experiment, we are able to predict very precisely the form of the interference pattern seen on the screen (even though we cannot predict where any particular atom will land). Much more impressive is that quantum mechanics predicts very precisely the properties of atoms and molecules and their constituent particles, as well as the nature of the forces that hold them together to give the richness and variety of structures we see around us. This predictive power is a sign of a successful scientific theory. What is so amazing is that we can do all this without knowing why we arrive at the

results we do. It seems we can manage perfectly well without having a picture in our heads of just how the atom gets through the two slits.

The majority of practising physicists have learned to use the theory without understanding why it works. In fact, some of the most prominent scientists of our age have admitted openly that no one really understands quantum mechanics! Should this not be a cause for concern? We will investigate in this chapter the differing attitudes and views that physicists have held, and still hold, regarding the issue of interpretation.

I am sure I will displease many physicists with what I will say in this chapter, since I plan to adopt an agnostic attitude towards the various views held. After all, my aim in this book is to explain what quantum mechanics is and why it is so strange. But it would be extremely dishonest and arrogant of me to pretend that all is rosy in the quantum garden. Many physicists, some of them close colleagues and research collaborators of mine, feel strongly that there is no problem. They would argue that drawing attention to the conflicts between different interpretations of what is, after all, a perfectly well understood, logically consistent, and successful mathematical theory is unnecessary and pointless.

But I would be equally dishonest, and probably make this book a less interesting read, if I did not climb off the fence on certain occasions to make my personal views clear.

Let me begin with a strong but nevertheless accurate statement: No one interpretation of the quantum formalism has been proven to be any better than the rest, other than on aesthetic grounds or personal taste. This has therefore suggested to many that it is a futile exercise to argue the relative merits and shortcomings of the various interpretations. Worse still, many believe that there is no true interpretation and that they are all equally valid ways of thinking about what is going on.

This view is embodied in the widely quoted 'shut up and calculate' interpretation, which suggests that since it has proven (so far at least) impossible to find the right interpretation, it is a waste of time talking about it. Let the philosophers worry about such issues while

physicists get on with using the quantum mechanical formalism to learn about nature.

For over half a century most serious physicists frowned upon issues of interpretation. They argued that quantum mechanics is unique among scientific theories in that, while it has tremendous predictive power, all that it can comment upon, by definition, are the results of measurements. That should be all we need concern ourselves with, and not worry about requiring a unique interpretation in order to make progress. Such a pragmatic, or 'instrumentalist', view is rooted in the philosophy of 'logical positivism' that happened to be popular in Europe at the time that quantum mechanics was born. Of course I do not want to stray from physics into philosophy, but I'll give the basic gist of this view: if two people have differing opinions but no way of settling their differences through empirical facts then their conflicting statements are meaningless and they should go and have a beer instead.

And my position on the matter? I subscribe to what I call the 'shut up while you calculate' view which leaves me free to contemplate the relative merits of the different interpretations of quantum mechanics when I am not thinking about Greek symbols, writing computer code, or scribbling algebra on my blackboard. Unfortunately, I have yet to settle on one interpretation after almost twenty years of worrying. I would say that quantum mechanics allows us to communicate with nature fluently by following well-understood rules of mathematical grammar, but we still lack a unique translation of what is being said!

In saying that, I must admit that like most other physicists I am burdened with the legacy of Bohr and what is regarded as the standard interpretation of quantum mechanics. It is the interpretation favoured in quantum textbooks and taught to physics undergraduates as though it were the only conceivable explanation of what is going on, although this attitude has been changing in recent years. In its favour is the fact that it is the simplest interpretation. In providing recipes for carrying out calculations in quantum mechanics, it is the ultimate tool of the pragmatist and is known simply as the Copenhagen

interpretation. Unfortunately, it has nothing to say about some of the deepest quantum mysteries and instead simply side-steps many of the issues. However, the name covers a whole spectrum of views, albeit with a common core.

To give you an example of how ingrained the Copenhagen view is in the teaching of the subject, many of the statements I made earlier in the book are ones that would not feature in some of the alternative, equally valid, interpretations. For instance: I went to great lengths to explain how the wavefunction is not a real physical entity but merely a set of numbers that allow us to make predictions about measurements. This is only according to the Copenhagen view and is not a feature shared by others you will meet, in which the wavefunction represents something physically real. Even more surprisingly, I need not have insisted in the two-slit trick that the atom must 'somehow' go through both slits at once. It turns out that such a statement is not forced upon us by either the quantum formalism or experimental observation. In the de Broglie-Bohmian interpretation we will see how it is perfectly reasonable to assume the atom only went through one or other of the slits, and yet we can still end up with the interference pattern!

First though, let us take a closer look at what Bohr, and the generations of physicists who claim to follow him, seemed to be insisting on.

The Copenhagen interpretation

The Copenhagen view is not so much an interpretation of a scientific theory as an ideology or a philosophical stance. Nor do all physicists agree on what it stands for. Indeed, it has been stated[1] that

'The Copenhagen interpretation is not a single, clear-cut, unambiguously defined set of ideas but rather a common denominator for a variety of related viewpoints.'

[1] Max Jammer, *The Philosophy of Quantum Mechanics* (Wiley & Sons, New York, 1974) p. 87.

The Copenhagen explanation: since we can never know what goes on behind the quantum curtain without affecting the results, it's a waste of time asking the question and we can only talk about what we *can* see.

It was developed through discussions that took place in the mid- and late-1920s between Niels Bohr and the group of brilliant young geniuses he had gathered around him at his then-new institute in Copenhagen. The most notable of these was Werner Heisenberg, whom you met earlier. Heisenberg's main contribution to physics was his development of an alternative formulation of quantum mechanics to Schrödinger's wave equation, known as matrix mechanics. I have avoided discussing this approach since it is highly mathematical, but many practising researchers in quantum physics tend to prefer it as a more elegant and powerful technique than Schrödinger's approach. Indeed, a combination of the two approaches is often necessary.

Others, such as Max Born, who was the first to suggest the probabilistic interpretation of the wavefunction, also contributed to the development of the Copenhagen view. Later on, prominent

physicists such as the American John Wheeler were to extend and clarify many of Bohr's ideas. But on the whole, the Copenhagen interpretation should be credited to Bohr himself.

It is remarkable that Bohr's views are still held in such high regard even today, three-quarters of a century after they were first formulated. This is best summarized by the physicist Roland Omnès (one of the developers of the idea of decoherence):[2]

'Ordinarily, a physical theory is presented in a sufficiently precise and consistent way so that one does not feel it necessary to quote the founders; rather one only strives to maintain their spirit and inspiration. Contrary to this sane custom, the most complete books devoted to the Copenhagen interpretation are all reprints of original articles or learned commentaries, becoming more and more commentaries upon commentaries as time went on. They devote much space to an endless discussion of the difficulties facing interpretation, often involving arguments where the philosophy of science becomes more important than physics itself.'

Of course this should in no way detract from Bohr's genius and insight, but the clarity of his ideas in his own writing left much to be desired. This led to a variety of interpretations of his interpretation... if you see what I mean.

There are, however, a number of key ideas common to the numerous, sometimes even contradictory, versions of the Copenhagen interpretation.

First and foremost, it states that we can never describe a quantum system independently of a measuring apparatus. It is a meaningless question to ask about the state of the system in the absence of the measuring device, since we can only ever learn something about the system if we take it in conjunction with the device we use to look at it.

2 Roland Omnès, *The Interpretation of Quantum Mechanics* (Princeton University Press, 1994), p.81.

Does a theory need an interpretation?

The notion that it is acceptable for more than one interpretation of a scientific theory to coexist, provided they do not make different predictions, is probably unique to quantum mechanics. An interesting parallel can be drawn by comparing the situation with that of another of 20th-century physics' great achievements.

In 1905, just a few months after publishing his Nobel Prize-winning paper on the photoelectric effect, Einstein completed the work he is best-known for: his special theory of relativity. This stated that different observers moving relative to each other would disagree about distances and time intervals between two events. But since no one can claim to be in a privileged reference frame then the notion of absolute lengths and times disappears. This could only be understood by unifying the concepts of space and time. Thus, two observers moving at high speed relative to each other will not agree on the length of an object. This is sort of balanced out by the fact that they will also not agree on the timing between two events as measured by their clocks.

Distances between two points and the rates that clocks tick as measured by different observers are related to each other through a set of algebraic formulas known as the Lorentz transformation equations, after the Dutchman Hendrik Lorentz who first wrote them down a year before Einstein published his work. Indeed, much of the groundwork for relativity theory had already been carried out before Einstein, and an early form of the equations had been independently proposed by Lorentz and the Irish physicist George Fitzgerald in the 1890s to explain a famous experiment that uncovered the property that light is able to travel through empty space.

The trouble was, Lorentz and Fitzgerald had the right equation, and got the right answer, but for the wrong reason. They misinterpreted what was going on by assuming that the apparatus measuring the light speed was squashed in length due to its motion through a mysterious 'ether' that pervades all space. Einstein's great achievement was to propose a simple postulate with which he was able to give the correct interpretation of the physics. The notion that light needs some kind of medium to carry it along – in the same way that a water wave needs water – was shown by Einstein to be unnecessary. Everything slotted into place when he made the bold suggestion that not only can a beam of light travel through empty space but that it would be measured by us to have the same speed no matter how fast we are moving relative to it.

This hard-to-believe yet crucial concept, and the ideas that followed in which space and time are unified, give us an understanding of why we got the experimental results we do. Thus, the Lorentz-Fitzgerald length contraction result turned out to be correct, in the sense that the numbers they calculated agreed with observation, but not for the reason they had proposed. Einstein gave the correct interpretation. We see clearly that the value of a good interpretation of a scientific theory brings us closer to the truth. Without a valid interpretation we would still be groping in the dark, no matter how well our theory agreed with experiment.

In quantum mechanics, no one has been able to come up with a unique and fully satisfactory interpretation. But does that mean that such a 'correct' interpretation does not exist?

Next, the role of the observer is central. Since we are free to choose what type of measurement to make (the position or momentum of a particle, the direction in which a photon is polarized or an electron is spinning) then the quantum entity cannot be said to even have these well-defined properties until we look. It must remain suspended in a superposition until we decide what we wish to measure. In this way, certain properties of the quantum system are only endowed with reality at the moment of measurement. Before that, they cannot even be said to exist in a definite classical sense until we measure them. Only the results of measurements are real!

Another tenet of the Copenhagen interpretation is that there has to be a clear demarcation between the (quantum) system being measured and the macroscopic measuring device (described by the laws of Newtonian, or classical, mechanics). Thus, while the latter is ultimately composed of atoms too, it must not be treated as being equally subject to the quantum rules. The act of measurement brings about a sudden jump in the state of the measured system from a combination of potential properties to one actual outcome. The notion of 'collapse of the wavefunction' upon measurement was first coined by Heisenberg in 1929.

It is not surprising that a theory as counter-intuitive as quantum mechanics requires desperate measures to make some sense of it. Maybe this is all we can say on the matter. I shall therefore point out my objections to the Copenhagen view to justify why I feel, along with many other physicists, that it has had its day.

I do not consider the Copenhagen interpretation to be a real interpretation at all. It is a set of rules to abide by in order for us to make use of the quantum formalism without worrying about its meaning. Thus, not only does the Copenhagen interpretation not explain how the atom goes through the two slits, it states categorically that even to ask such a question is meaningless and we should restrict our comments to the interference pattern on the screen (the measurement). The Copenhagen interpretation succeeds in banishing logical contradictions and inconsistencies by only allowing those questions to be posed which concern the results of measurements.

Many of the founders of the Copenhagen interpretation, such as Bohr, Heisenberg, and Wolfgang Pauli, became somewhat contemptuous of a number of later attempts to work out a physical picture of the quantum world. They saw this as a futile attempt to return to an outmoded (Newtonian) way of thinking that had been replaced for good. Indeed, later in life, Bohr even refused to look at alternative interpretations, presumably in the hope that they would just go away. Likewise, Heisenberg was dismissive of such programmes, since

'they only repeat the Copenhagen interpretation in a different language [and] from a strictly positivistic standpoint one may even say that we are here concerned not with counter proposals to the Copenhagen interpretation but with its exact repetition in a different language.'[3]

He felt that the formalism of quantum mechanics determined a unique interpretation, and alternative approaches differed only in the way they interpreted the mathematics rather than implying any different physics. Needless to say I disagree, since to hold such a view, one must first subscribe to the Copenhagen school of positivism!

My second criticism is that the Copenhagen interpretation says nothing about how the process of wavefunction collapse takes place. A strange divide exists in time – before and after measurements – for the quantum system being measured. Before we look, it evolves according to the Schrödinger equation. This evolution – the way the wavefunction changes over time – involves no uncertainties or probabilities. But as soon as a measurement takes place we must abide by a different set of rules which bring in the quantum probabilities. This distinction between an ordinary physical process and a measurement is not something the Copenhagen view can, or even attempts to, explain.

[3] Werner Heisenberg, *Physics and Philosophy* (New York: Harper and Row, 1958), p. 129.

Finally, and most importantly of all from a personal point of view, by giving the observer such a privileged status, the Copenhagen interpretation denies the existence of an objective reality that exists in the absence of observation.

Copenhagenists often accuse those who still search for interpretations of the quantum formalism as being lost in metaphysics, rather than real physics. But if there does exist a 'correct' interpretation of quantum mechanics then looking for it is an attempt to explain how nature really behaves. And to be satisfied with a set of ad hoc rules to abide by in order to carry out computations can only ever be a temporary measure.

So what else is there on the market? And if other interpretations are able to give a more rational and convincing explanation, why is the Copenhagen interpretation still favoured by the majority of physicists?

The de Broglie-Bohm interpretation

Of the founding fathers of quantum mechanics, several were dissatisfied with what they perceived as the prevailing Copenhagen dogma. From the very start, Einstein, Schrödinger, and de Broglie all tried hard to put forward opposing interpretations of quantum mechanics, but they were no match for the Copenhagen group's talent, drive, and sheer persuasiveness. It is well-documented that Bohr and Einstein had numerous professional arguments over the meaning of it all, and it is acknowledged that Bohr finally won the day. But the reason for this was that Einstein had been insisting on a way of banishing all quantum weirdness. We now know this is not possible, but it does not mean Bohr was right!

In fact, it was Louis de Broglie who came up with the first serious alternative to the Copenhagen view. He called it his 'principle of the double solution', implying a synthesis of the wave and particle aspects of matter which he had proposed. The Solvay conference at Brussels in 1927 represents one of the landmark events in the development of quantum mechanics. It has been said of this meeting that never has

so much been clarified by so few in such a short space of time. All the major players in theoretical physics at the time were in attendance, including Einstein and Bohr.

De Broglie presented a paper in which he hypothesized that the wavefunction might actually represent a real physical wave that guided the real quantum particles along certain paths. He had not had the time to get all his ideas clear in his head, and so was not in a position to defend them in the face of strong criticism from some of the delegates at the conference. In fact, with only two years having elapsed since Heisenberg had published his matrix mechanics, the Copenhagen circle was completely dismissive of any approach that attempted to revert back to 'realism' as being too pictorial and quite wrong in the quantum domain.

The de Broglie-Bohm explanation: the atom remains as a localized classical particle throughout and, as such, only ever goes through one of the slits. However, the quantum potential's influence is spread over both slits and it guides the atom along a definite yet uncontrollable and unpredictable path etched out by the pattern of its distribution.

A quarter of a century later they were shown to have been too hasty, even dogmatic, in their attitude.

David Bohm, an American physicist who settled in England after being forced out of his job at Princeton for supposedly un-American activities during the McCarthy era, succeeded in doing something many had believed impossible. In 1952, he published two papers which were essentially an extension of de Broglie's original idea. With this work, he showed that the Copenhagen interpretation was not the only one compatible with all experimental results.

It is important to stress that Bohm's interpretation does not require any additions made to the equations of quantum mechanics. It was not the theory – the mathematical framework from which we calculate the physical properties of quantum systems – that people felt uncomfortable with, but its meaning. So before I say a little more about Bohm's idea, I must stress again that these two interpretations (Copenhagen and de Broglie-Bohm, for de Broglie seems not to have received the credit he deserved) do not make any different measurable predictions about the subatomic world. They can't do if they use the same theory. What they do give us are two very different views on what is going on in the two-slit trick. They therefore differ in the underlying physics and not just the language used to describe the formalism.

Bohm suggested that the wavefunction was not simply a mathematical entity but a real physical presence. He followed de Broglie's old idea of recasting Schrödinger's equation in a form that led to the picture of definite particles being guided by some sort of force field.

Fans of the de Broglie-Bohm approach point out that what it gives back to us is nothing less than reality itself. We are again able to ask what is really going on when the atom confronts the two slits. We may not like what we get, and some physicists use this as an excuse to attack this interpretation, since quantum weirdness – such as the nonlocal nature of the wavefunction we met earlier – is no longer buried in cosy mathematics but is 'in your face'. Bohm allows us to ask

questions about how nature is behaving when we are not looking. To many physicists this seems a pointless exercise since we cannot judge the truth of our statements until we do look, and then we might as well choose the Copenhagen interpretation since it is a minimalist interpretation concerning only the results of measurements.

So how does the de Broglie-Bohm interpretation explain the two-slit trick? Well, it claims that part of the information buried within Schrödinger's wavefunction describes a form of quantum energy, called the quantum potential, which propagates throughout space and guides the quantum particles along certain paths. So, while the atom itself will only ever go through one slit or the other, the quantum potential actually permeates both. The pattern that builds up on the screen is now due to the way the quantum potential behaves on the other side of the slits. The shape of the quantum potential can be thought of as causing each atom to move along a definite, but unpredictable, path. It is only when many atoms have gone through that the true shape of the quantum potential emerges in the pattern of hits on the back screen.

The interference pattern disappears when we try to spy on the atoms because we disturb the quantum potential. In the same way, closing one of the slits alters the quantum potential everywhere, causing a change in the possible trajectories of the atoms going through the other slit.

Disturbing the quantum potential in one location instantly modifies it throughout the whole of space! And while this nonlocality cannot be used for faster than light signalling, due to the built-in unpredictability of how it will affect any given particle it guides, it is still this aspect of the de Broglie-Bohm view that most people find distasteful.

What we must not forget is that quantum mechanics is nonlocal whatever interpretation we choose to apply. It is just that, in the Copenhagen view, the nonlocality is just another mathematical property of the wavefunction, and hence deemed not to be 'real'. The price to pay is that certain properties of the atom are also not real (until measured of course). The physicist John Bell, who was a keen

supporter of Bohm's ideas, always stated that he would much rather give up locality than reality itself!

In the de Broglie-Bohm interpretation we no longer have to think of atoms as spread-out when they go through the two slits; they are definite, localized particles at all times. But it is important to point out that we can still only calculate probabilities for finding the atom in a certain location, using the wavefunction in the same way as before. So despite it restoring the old idea of determinism to nature, it is not a return to Newton's clockwork universe. It is fundamentally impossible for us to ever control the initial conditions for any given particle such that we are able to predict its definite trajectory, since any such attempt will cause a change to the quantum potential. Put another way, we can never know before hand which slit a given atom will go through.

Also, while this interpretation assumes that each particle follows a real, definite trajectory, it is impossible for us to prove this.

So why has the de Broglie-Bohm interpretation not caught on? Its supporters still represent a minority among physicists. The usual accusation is that the de Broglie-Bohm interpretation is only successful because it is based on the same mathematical theory as the Copenhagen interpretation. Anything beyond that is observationally meaningless and solely based on classical prejudice. But even Heisenberg himself could think of no stronger criticism than to say that it was 'metaphysical' and 'ideological'.

Other physicists find the extra assumptions about the existence of the quantum potential unnecessary, and are uncomfortable with its, very real, nonlocality. For instance, in the EPR experiment I discussed in Chapter 4, one of the two particles would instantaneously influence its distant partner via their combined quantum potential. But Einstein has taught us that faster-than-light communication is impossible. So if the quantum potential is real then quantum mechanics must conflict with relativity theory. But, as mentioned earlier, in practice things are not as bad as this – even if there is a quantum potential, such faster-than-light signalling between the particles can never be utilized to send messages, due to the built-in

probabilistic (that is, unpredictable) nature of individual quantum measurements.

Possibly the real reason that this version of quantum mechanics has not become the standard one is simply due to historical contingency. This was the view of American physicist Jim Cushing who argued that the real reason the Copenhagen interpretation is still regarded as the standard one is because it came along first and was advocated by stronger personalities. Its opponents in the 1920s (Einstein, Schrödinger and de Broglie) were not united in their criticisms of it. Thus, as Cushing puts it[4]

'The Copenhagen interpretation got to the top of the hill first, and to most practising scientists, there seems to be no point in dislodging it.'

Bohm's programme actually belongs to a class of interpretations known collectively as 'hidden variables' theories, and is the most sophisticated of these approaches. These hidden variables represent a deeper level of physical reality that is hidden from us but is the origin of quantum uncertainty and fuzziness. In the de Broglie-Bohm interpretation, the hidden variables are the definite positions of the particles. Thus, it differs from the Copenhagen view when it states that a quantum particle does have definite position and momentum prior to measurement but we are unable to access these properties. The uncertainty principle becomes now just a statement about our inescapable ignorance.

The mathematician John von Neumann appeared to have provided a mathematical proof in 1932 that ruled out the possibility of hidden variables interpretations, lending support to the Copenhagen view that certain properties of a quantum object cannot be said to exist until we measure them. It was only many years later that physicists discovered that von Neumann's

[4] James T Cushing, *Quantum Mechanics: Historical Contingency and the Copenhagen Hegemony* (University of Chicago Press, 1994).

'proof' was wrong.[5] Provided you allowed for nonlocality, there was no reason why you couldn't have an interpretation based on objective reality such as that of Bohm. Attempts to rule out all hidden variables interpretations are known as 'impossibility proofs'. John Bell stated that 'what is proved by impossibility proofs is a lack of imagination'. More recently, however, David Mermin, a physicist who has contributed much to advancing Bell's ideas, hit back with the retort that 'what is proved by possibility proofs is an excess of imagination'.

The many-worlds interpretation

Several years after Bohm's work, another American, by the name of Hugh Everett III, proposed what he called the 'relative state interpretation' (sounds pretty innocuous doesn't it?). This has since come to be regarded as both the most outlandish and extravagant interpretation of quantum mechanics and yet the most straightforward, depending on which side you are on. In fact, I tend to oscillate between these two extremes myself, one day wondering how anyone could be so silly to even give it the time of day, another wondering how anyone could conceivably consider any alternative.

Everett's interpretation, like the de Broglie-Bohm one, is free of the measurement problem that plagues the Copenhagen version. Nor does it require the physical nonlocality of Bohm. And while decoherence fits in comfortably enough in the first two interpretations, it seems (to me anyway) to find a home most naturally in Everett's approach or its modern-day variants. Well, so far so good it seems. But never forget that quantum weirdness must show up in some guise or another. Here, it is in the outrageous requirement that we must inhabit one of an infinite number of parallel universes!

[5] Actually, von Neumann was shown to be wrong just three years later by Grete Hermann, but the Copenhagen view was so ingrained that her proof seems to have been completely ignored.

The many-worlds explanation: all possible realities co-exist. The atom goes through a different slit in each universe and the two universes overlap only at the level of the single atom. In each universe, the atom feels the presence of its parallel self which has gone through the other slit. The superposition, and hence interference, is the result of superposition of universes.

What could possibly lead to such a conclusion? It is as though the only certain criterion for a valid interpretation of quantum mechanics is that it must sound weird. Maybe there is another interpretation out there whereby all weirdness is banished apart from the notion that quantum measurement should involve signalling backwards in time. Oh, hang on a moment, that one has already been discovered! It is called the transactional interpretation and I'll say something about it towards the end of the chapter. But you can see what I am getting at. These differing yet quite serious views of the way nature conducts business on the microscale certainly make all the New Age pseudoscientific theories, such as the mystical powers of crystals and pyramids, seem a little dull in comparison.

Back to Everett. His interpretation has since spawned a number of variants. While his original idea is now known as the many-worlds interpretation, there is also the multiverse interpretation, the many-histories interpretation and the many-minds interpretation. I quite like the first, do not really understand the second, and am not at all keen on the third.

The basic idea is the following: When a quantum system is faced with a choice of alternatives such as a particle going through one of two or more slits then, rather than the wavefunction entering a superposition, we think of it, and the whole Universe along with it, as splitting into a number of realities equal to the number of options available. These different worlds/universes/branches will be identical to each other apart from the different option chosen by the particle: in one universe it has gone through the upper slit, in the other it has gone through the lower slit. The universes overlap, only in that region where interference is taking place, until decoherence sets in. This then causes them to separate into non-interacting independent realities. That is it. There is no measurement process any more and the wavefunction never needs to 'collapse'. Schrödinger's cat will end up dead in one universe and alive in the other. We, as observers, will also split and so only ever see the outcome of our branch. But there will be other copies of us in parallel universes for whom the alternative outcomes are realized.

Hugh Everett became disillusioned and discouraged by the lack of support for his views from other physicists – indeed, it is said that Bohr gave him the brush-off when Everett visited him in Copenhagen. He left physics to work first as a defence analyst, then as a private contractor to the US defence industry, where he made a fortune working out how to maximize kill rates during a nuclear war. Nice. Interest in his work was revived in the late 1960s by Bryce DeWitt who coined the term 'many-worlds interpretation'.

Oxford physicist David Deutsch, one of the founding fathers of the field of quantum computing, proposed a variation on Everett's idea in which all possible universes already exist and don't have to

wait until faced with quantum alternatives to split. Deutsch has also proposed a test of his interpretation, which requires the development of artificial intelligence. The test would be to see if wavefunction collapse takes place or not, since in the many-worlds interpretation it does not.

What else is on the market?

Some interpretations have, over the years, been discredited or dismissed, either as being too unlikely or because they were ruled out by new experimental results. For instance, relatives of the de Broglie-Bohm approach, known as local realism theories, were scratched off the list following the experimental results of Alain Aspect in the early 1980s which confirmed the violation of Bell's inequality (discussed in Chapter 4). Similarly, hardly anyone still takes seriously the notion of consciousness being a requirement for collapse of the wavefunction.

One of the early approaches was the statistical interpretation – interestingly the one favoured by Einstein – which assumes that quantum mechanics only makes statements about whole collections of measurements (of identical quantum systems) rather than on an individual measurement, hence the reference to the overall 'statistics'. Since experiments can now be performed on individual systems, even individual atoms, this interpretation requires major clarification by its small group of supporters if it is to survive.

Two recent additions to the list are the transactional and the consistent histories interpretations. The first, due to John Cramer, is similar to the de Broglie-Bohm view in that it is explicitly nonlocal. Indeed the nonlocality is even more severe here: it is not instantaneous connections across space that are required but connections across time! The act of opening Schrödinger's box sends a signal into the past to tell the radioactive nucleus to decay or not.

The consistent histories approach, mainly due to the particle physicist and Nobel Prize winner Murray Gell-Mann and his

Left top: the transactional explanation: while this interpretation does not help with explaining the central mystery of how the atom manages to go through both slits at once, it does attempt to explain how the atom knows before it goes through the slits whether or not it is being watched. On arriving at the screen, or in a detector close to one of the slits, the atom sends a signal backwards in time that tells the advancing wave how to behave: either to go through both slits and interfere or to go through only one of the slits.

Left below: the sum-over-histories explanation: the atom remains as a particle but it explores all possible paths simultaneously, however unlikely their route. Summed together, all paths cancel each other out leaving just the physical path taken by the atom. But the way the paths cancel depends on what options are available; if both slits are open, more paths are available and the cancellation is different.

collaborator James Hartle, has yet to really catch on, but is gathering a growing band of supporters. It combines wavefunctions and probabilities in a consistent way that does not require acts of measurement. It was first proposed by Robert Griffiths in 1984, and further developed a few years later by Roland Omnès. A 'history' according to this view is defined as a sequence of quantum events at successive times. The neat thing about this approach is that it allows us to assign probabilities for different events – such as when a radioactive atom will decay – even if they are out in interstellar space far from any measuring device.

I should also mention that there are several interpretations based on dynamical reduction theory.[6] These approaches require an extra something that causes the spontaneous collapse of the wavefunction all by itself (in the absence of measurements) every now and again. While the theory can be made to agree with all known observations, the real physical mechanism that would bring about this collapse is still unknown. Experiments in the near future should be able to check whether this does indeed take place. If so then quantum mechanics will need modifying. Most physicists think this highly unlikely.

[6] The best known is due to three Italian physicists: Ghirardi, Rimini and Weber, proposed in 1986 and referred to as the GRW approach. Another was proposed by Roger Penrose who suggests that gravity can do the job of collapsing the wavefunction.

Quantum Reality, According to de Broglie and Bohm

Chris Dewdney School of Environmental and Earth Sciences, University of Portsmouth

Even though the de Broglie-Bohm interpretation is not the most popular one among physicists, a number of different versions of it exist. The common basis of the different approaches is that particles have well-defined positions that evolve according to deterministic equations of motion. Given some assumed initial positions of all the particles in a system (remember these cannot be controlled – see main text) and the way in which the wave function changes with time, one can calculate exactly how the particles will move. So the whole future (and the past) of the system can be predicted, including the results of any measurements that may be involved. In this respect there is no more mystery to quantum mechanics than there is to classical mechanics. The different approaches arise from what we say about how the trajectories come about.

Bohm originally presented the theory in terms of a new type of force, arising from a quantum potential, which pushes the particles to the right places and accounts for the difference between quantum and classical behaviour. Bohm wanted to bring out, by contrast, the essential differences between the quantum and the classical worlds. Others follow John Bell's version and argue that all we need is the equation that determines the trajectories (which should be seen as a new type of law of nature in itself) and that we should eschew

all talk of quantum potentials and quantum forces that, regrettably, straitjacket the theory in an ugly Newtonian form. One of the problems, which may have hindered the acceptance of the de Broglie-Bohm interpretation in the past, is that, in all but the most simple and uninteresting cases, the trajectories cannot be calculated by hand. However, since the late 1970s, starting with the two-slit experiment, many explicit calculations have been carried out clarifying exactly how the approach works.

An interesting feature of the de Broglie-Bohm interpretation, as Bohm pointed out in 1952, is that, although the basic theory does reproduce all of the expected results, one can consider variations, which arise from relaxing some of the assumptions of the interpretation, that introduce novel observable effects. Work is currently underway to determine the circumstances in which such effects may be detectable.

One of the most fundamental features of quantum mechanics that is put into stark relief by the de Broglie-Bohm interpretation is nonlocality. Nonlocality is the coupling of separate systems in circumstances in which, according to relativity, no physical interaction between them should be possible. According to relativity all influences must propagate at, or less than, the speed of light, but nonlocality defies this edict. One of the most interesting questions in de Broglie-Bohm theory today is how its explicit nonlocality can be squared with the requirements of relativity theory. It is accepted that no measurements that can be done with quantum systems can reveal a conflict with relativity by demonstrating superluminal influences explicitly: we can't use quantum mechanics to signal at speeds faster than the speed of light. It is also accepted that any realist interpretation of quantum mechanics must have nonlocal processes. Bohm argued that, even though his interpretation gives rise to the statistical results of quantum theory, that are known to accord with relativity, at the fundamental level of the individual processes (which account for these results) the spirit of relativity is violated. Superluminal influences are to be allowed and furthermore, he argued, contrary to the basic principle of relativity, there exists

a special, or preferred, frame-of-reference in which the influences are instantaneous. Many physicists cannot contemplate such a departure from relativity, even at the level of hidden processes, and cite this as a reason for dissatisfaction with Bohm's approach. Experiments have even been carried out to try to find out which feature of an experiment fixes the preferred frame in which the influences act instantaneously. But Bohm's approach is not the only option.

More recently, various proposals have been made that show how the de Broglie-Bohm approach can be extended, without the need to introduce a preferred frame-of-reference, by embedding nonlocality in a thoroughly relativistic de Broglie-Bohm interpretation. This work shows that the major objection, that the de Broglie-Bohm interpretation contradicts relativity theory, can be overcome. Of course there is still much work to be done, especially concerning the processes described by relativistic quantum field theory. However, the fact is that there is no knock-down argument which disqualifies the de Broglie-Bohm interpretation, and ultimately every objection to the approach is based on a predilection for certain sorts of ideas rather than others.

Where do we stand?

It is really only in the last couple of decades that the problems of interpretation have come to the fore of respectable research. This has been partly due to the new generation of incredibly sophisticated and novel experiments in atomic physics and optics, as well as the remarkable work being done in the new fields of quantum cryptography and quantum computing. In many of these studies physicists are manipulating individual atoms!

As a student I was never alerted to the problems of the interpretation. And while the topic of this chapter only serves to confuse students and distract them from being able to do useful and meaningful calculations and measurements, it does seem as though discussions surrounding such issues were, until recently, regarded as taboo. Murray Gell-Mann put it thus:

> 'Niels Bohr brainwashed a whole generation of physicists into believing that the problem had been solved.'

Even in Bohr's own words:

> 'there is no quantum world. There is only abstract quantum physical description. It is wrong to think that the task of physics is to find out how nature is. Physics [only] concerns what we can say about nature.'

I disagree. At least here I feel Einstein was right. He believed that the job of physical theories is to 'approximate as closely as possible to the truth of physical reality'. So in true *X-Files* spirit, I would prefer to think that 'the truth is out there'. Whether we will ever reach it or not I do not know but I am confident it will not turn out be a futile search. Just because the formalism of quantum mechanics allows us the luxury of several interpretations that we cannot (yet) choose between does not imply that there isn't a correct interpretation. Of course, we may never be able to find it, but it is too arrogant of us to claim that since we are unable to choose, then neither does nature.

Chapter 7
The Subatomic World

Having read this far, you might have got the impression that quantum physicists spend their time debating the nature of reality and arguing about such deep and meaningful issues as the definition of the words 'event' and 'phenomenon', and what it means to measure something, or thinking up ever-more-imaginative ways of describing what might or might not be going on in the microscopic world when we are not looking. This is far from the truth. The majority of physicists could not care less about the issues surrounding the interpretation of quantum mechanics, and rightly so. They are too busy using the theory to understand the structure and properties of the subatomic world.

Indeed, without this 'shut up and calculate' attitude, we would not have made the tremendous advances over the past half century in science and technology that I will discuss in Chapter 9.

Now that I have exhausted the debate about how we might unravel the mystery of the two-slit trick, we can move on to consider just how quantum mechanics contributed to the birth of new fields of scientific endeavour in the search for the ultimate building blocks of matter, and how these building blocks interact and combine to produce the beautiful complexity of the world around us. Throughout the last century, physicists probed ever deeper and ever smaller, first looking inside the atom, then inside its nucleus,

and later inside the particles that make up the nucleus. The search for the ultimate constituents of matter has been likened to peeling an onion. As each layer of the onion was peeled back, a more fundamental structure was revealed beneath. This chapter is therefore the story of the development of atomic, nuclear and particle physics, and how quantum mechanics has been the guiding light throughout this odyssey of discovery. Of course, it is impossible for me to do justice to the history of these fields in just one chapter, so this will only be a whistle-stop tour, pausing briefly along the way to take a closer look at some of the highlights and the personalities who have shaped so much of 20th-century physics.

More importantly, we have yet to reach the end of the quantum weirdness list. There are still several nuggets, such as quantum tunnelling, spin and Pauli's exclusion principle, to keep us going.

Mystery rays everywhere

I described in Chapter 2 how the years between 1895 and 1897 marked the birth of what is referred to as modern physics. Looking back today, those first few exciting years might more appropriately be called the 'mystery rays period'. They were being discovered all over the place. The run-up to the new century saw the discovery of X-rays, radioactivity, and the electron, each of which came as a complete surprise to the scientific community.[1] Within the next decade they were to bring Nobel Prizes to their discoverers: Wilhelm Röntgen, Henri Becquerel and Joseph (J.J.) Thomson, respectively. Remember how, in Chapter 2, we traced the birth and development of quantum theory following Planck's revolutionary idea at the turn of the 20th century. At the time, however, his work was far from being the most exciting and interesting area of research in physics. Instead, it was the discovery of X-rays that had captured the imagination of the scientific community. Here were invisible rays that could penetrate

[1] Well, maybe the electron wasn't a complete surprise.

through solid matter, forming an image on photographic film on the other side. This amazing property was an instant hit with the general public around the world, and many soon saw its remarkable benefits in medicine and industry.

Soon after this discovery, Henri Becquerel, who was also interested in the origin of X-rays, studied a uranium compound that gave off a fluorescent glow and led to results similar to those seen by Röntgen in his X-ray experiment. Becquerel found the radiation from uranium could also pass through solid matter, such as the black paper wrapping he was using to protect his photographic film from sunlight, and leave its traces on the undeveloped film. His initial thought was that the uranium was also emitting X-rays as part of the fluorescence brought about by exposure to sunlight. But he soon discovered that whatever was being emitted had nothing to do with sunlight or X-rays. Two years later, Marie Curie, who together with her husband Pierre was now also working on the mystery rays, coined the term radioactivity.

Meanwhile, one of the giants of experimental physics, the Englishman J.J. Thomson, discovered the nature of yet another kind of ray. It had been known for many years that an electrically charged metal plate inside a vacuum tube emitted what was dubbed a cathode ray. But no one knew what it was made up of. Thomson showed that it consisted of negatively charged particles that were much smaller than atoms.

In a manner reminiscent of Planck's idea of energy quantization, but preceding it by a decade, the Irishman George Stoney had suggested that electricity might not be continuous, but come in tiny indivisible lumps he called 'electrons'. Soon after Thomson's discovery, Lorentz suggested these must be Stoney's electrons. Despite his initial objections to the idea, it was Thomson who received the Nobel Prize for the discovery of the electron, the first elementary particle. This is often the way discoveries are credited in science; Thomson did not discover cathode rays, nor did he even name them. But he did get the Nobel Prize for the experiment that showed what they were. It was a discovery that marked the birth of subatomic physics.

Into the atom

So where did things stand at the start of the last century? Thomson had suggested that electrons were part of the internal structure of atoms. But since they were negatively charged and atoms were neutral, this implied that atoms must also contain a positive charge to cancel out the charge of the electrons. Thomson therefore proposed the first tentative atomic model, in which electrons were evenly distributed throughout a sphere of positively charged atomic 'stuff'. The size of the sphere could be reliably calculated from quantities known at the time. This became known as Thomson's 'plum-pudding' model of the atom, in which the electrons were the plums. At the same time, Becquerel and the Curies were convinced that radioactivity was also being emitted from within certain kinds of atoms. So while there were still physicists and chemists who did not even believe in the existence of atoms, others were already getting a glimpse of what lay inside them.

Born in New Zealand, Ernest Rutherford was destined to become one of the most influential figures of 20th-century science. He had arrived in Cambridge to work with Thomson and soon became fascinated by the new topic of radioactivity. He discovered that there were three types of radioactivity. One, which he called beta rays, turned out to be just Thomson's electrons. Another type, called alpha rays, appeared to consist of much heavier, positively charged, particles later shown by Rutherford to be helium ions (atoms of helium with their electrons stripped off). He also demonstrated that a third, electrically neutral, type of radioactivity, dicovered in 1900 by Paul Villard, was just a form of electromagnetic rays, like X-rays,[2] which he called gamma rays. These days of course, we refer to the alpha and beta types as particles rather than rays.

[2] Confirmation that X-rays were actually electromagnetic radiation the same as light had to wait until 1912.

During the first decade of the century, Rutherford was able to show that the Earth's crust had to be billions of years old. He did this by measuring the amount of helium trapped inside rock samples, in which tiny amounts of uranium ore had slowly been emitting alpha particles ever since the rocks had been formed. Each alpha particle would be trapped by the rock and would quickly acquire a couple of electrons to become a helium atom. Such simple yet irrefutable proof one hundred years ago that our planet had to be more than a billion years old is something which those who subscribe to the notion of creationism are unable to challenge with any credibility.

A few years later, Rutherford was to become the very first scientist to achieve the alchemists' dream of transmutation, when he carried out an experiment in which he transformed one element into another. This did not come as a surprise since he had, together with Frederick Soddy, already identified radioactive decay as a natural form of transmutation.

With the discovery of alpha particles, Rutherford was quick to realize that here was a tool for studying atomic structure. In 1911, two of his assistants, Hans Geiger and Ernest Marsden, carried out a series of painstaking experiments, in which a beam of alpha particles emitted from a radioactive source was aimed at a very thin gold foil. The scattered particles were detected as tiny flashes of light when they hit a photosensitive screen. Despite the foil being several thousand atoms thick, most of the alpha particles were found to go straight through with hardly any deflection – clearly, atoms must consist of mostly empty space. What was much more incredible was the fact that about one in eight thousand of the alpha particles would bounce right back again. This is impossible if we think of atoms in terms of Thomson's plum-pudding model.

Rutherford realized the significance of this result and came up with a much better model for what an atom should look like. Firstly, he knew that electrons were several thousand times lighter than alpha particles and so would not be able to deflect them from their paths. It seemed that the positive alphas were being deflected by the electrical repulsion of the positive charge within the atom. The

According to Rutherford's theory, the closer the alpha particle came to a head-on collision with the tiny atomic nucleus, the greater the angle of deflection. Geiger and Marsden's experimental results confirmed this picture beautifully.

only way to explain what was seen was if all this positive charge was concentrated in a very tiny volume to maximize its effect. Most of the alphas would miss it completely, while a very few would get so close to it that they felt the full force of the gold atoms' positive charge and got repelled back.

Rutherford named this tiny concentration of positive charge the atomic 'nucleus' and proposed a new model of the atom that consisted almost entirely of empty space, with a tiny positive nucleus containing almost all the atom's mass, surrounded by even tinier orbiting electrons.

It was at this point, in 1912, that the young Niels Bohr came to work with Rutherford and first applied Planck's quantum theory to the planetary model to explain its stability, a story I discussed in Chapter 2. But remember that the full quantum mechanics was still a decade away from being developed. In Bohr's model the electrons were still regarded as tiny 'classical' particles whizzing around the nucleus along fixed orbits. The advances made during the 1920s by

Cross sectional views of electron probability clouds.

Left: In a hydrogen atom, the single electron is in its lowest energy state and is most likely to be near the centre of the atom.

Centre: If the electron in hydrogen is excited to the next energy level its cloud changes suddenly. Now it has a small probability of being found in the centre but is more likely to be within a spherical shell some way out. Between the centre and shell, there is zero probability of finding the electron.

Right: A carbon atom containing six electrons. Four have no orbital angular momentum and their probability distributions are symmetric. The other two have a small amount of angular momentum, which means that the probability distribution of each could be in any one of three possible orientations – only one is shown.

Opposite: Electron clouds for a variety of atoms (from top to bottom in order of increasing number of electrons: hydrogen, silicon, iron, and silver). As each orbital is filled, further electrons must occupy higher energy orbitals.

the likes of Heisenberg and Schrödinger showed that such a picture of the atom was not only naïve but wrong in many respects. They showed that the quantum rules that Bohr insisted had to be obeyed by the electrons were more correctly associated with electron wavefunctions. By solving Schrödinger's equation, physicists could explain how the electrons arranged themselves in the atom. And it was only with hindsight that Bohr's ad hoc formulae could be seen to emerge from the complete quantum mechanics. The difference now was that each electron could not be thought of as a localized particle orbiting the nucleus, but instead had to be described by a

wavefunction which carried certain labels, called quantum numbers, that defined that electron's energy and the way it moves around the atom. Electron wavefunctions occupy the whole volume of the atom and provide us with a probability distribution that maps out where the electron is likely to be found if we were to 'look there'. You might have come across phrases such as 'electron cloud' or electron 'probability density'. This simply refers to this three-dimensional pattern of probability that does not change with time unless the electron changes from one quantum state to another (equivalent to the old idea of an electron 'jumping' from one fixed orbit to another)

if it loses or gains a quantum of energy or angular momentum.[3] Any such change will alter its wavefunction and the shape of the associated probability distribution.

So just as we have been forced to say that the atom goes though both slits at once, we must regard every electron as spread throughout the volume of its atom. This is the only valid picture of the atom that quantum mechanics allows us to have. The different regions within the atom where each electron prefers to spend its time depend on the shape of its wavefunction, which in turn is set by the quantum numbers for that electron. It's all very mathematical and counter-intuitive, which is why it is so much easier to stick with the Rutherford/Bohr planetary model of atoms. It may not be right but at least we can visualize it. But the whole of modern chemistry and much of physics is founded on this quantum structure of atoms. The quantum numbers the electrons have define how they arrange themselves in what are called 'quantum orbitals' (that are like fuzzy, strange-shaped orbits)[4] or 'energy levels' that can explain beautifully how all the elements are classified according to their chemical properties and arranged in the periodic table.

And quantum mechanics predicts much more than just the electronic arrangements in atoms. It also provides predictions about the electron transitions between different energy levels. This takes place for instance when atoms interact with light. Electrons can absorb photons provided the photon energies correspond to the energy gaps between the different levels. If this happens, the photon ceases to exist since it is entirely absorbed as pure energy to 'excite' the electron to a higher energy level. The electron does not often feel comfortable with this extra energy and soon emits a photon of exactly the right energy to allow it to drop back down to its original

[3] Remember that a 'quantum' here means the smallest lump or unit of a quantity that we consider as continuous in the macroworld, such as energy.

[4] The shape of an electron orbital is, as ever, something we can work out from the wavefunction; namely, the probability cloud showing the likely distribution of the electron around the atom

Quantum spin

Almost all non-technical books on quantum mechanics choose to explain the origin of quantum weirdness using the concept of 'spin'. I have tried to avoid it thus far despite it being probably the most 'quantumy' of all quantum properties. It is very far removed from anything we can conceptualize using everyday language.

In 1896, the Dutchman Pieter Zeeman discovered that when atoms were placed in a magnetic field and then excited, the spectral lines – the narrow bands of light forming a pattern on a screen produced by their emitted light – would split into several components. In certain cases, this splitting could be understood in terms of a classical (non-quantum) theory developed by Lorentz, but in general the effect could not be understood and was referred to as the 'anomalous Zeeman effect'. It wasn't until 1925 that Sam Goudsmit and George Uhlenbeck suggested that it might be to do with another property of the

atomic electrons that the magnetic field was bringing to the fore. Using concepts based on the old quantum theory involving Bohr's electron orbits, they suggested that, in addition to an electron's orbital motion around the nucleus, it was also spinning about its axis (just as the Earth rotates as it orbits the Sun). But this 'quantum spin' is unlike anything we could visualize based on our everyday experience of spinning objects like cricket balls or baseballs.

The problem was that, like the electrons' orbital angular momentum, this spin angular momentum had also to be quantized. Firstly, all electrons spin at exactly the same 'rate', and can never slow down or speed up. Their direction of spin is even weirder; it is a superposition of different directions at once, until we look. By doing so we must specify an axis (basically a direction) about which they are spinning, in which case we always find them spinning either clockwise or anticlockwise

The belt trick: As far as magic tricks go this one is pretty unimpressive, but it serves as a nice example of the nature of quantum spin. 'Fermion' particles, such as electrons, are characterized by having 1/2-integer spin. This means that rotating an electron by a full 360 degrees does not bring it back to its original state. We have to rotate it by a further 360 degrees to do that. I saw the following example described in a lecture by Roger Penrose*: Secure one end of a belt under a heavy book on the edge of a table. The idea is to twist the belt round and then remove the twist by passing the free end through a loop. You would have thought that a full twist could be cancelled by a single looping procedure. It isn't. However, if you give the belt two 360-degree twists then looping will completely untwist it. Don't read too much into this analogy, but it might serve to highlight how an electron must also make two revolutions to complete a 'full cycle'.

* The idea first appeared in a classic textbook called *Gravitation* by Misner, Thorne and Wheeler (WH Freeman & Co, 1973).

around that axis. Before we look, however, they will be in a superposition of spinning in both directions at once!

Finally, what we regard as a 360° rotation of the electron will not take it back to its original quantum state; it needs to make two full rotations to do that! The reason this sounds strange is that we cannot help but think of an electron as a tiny ball having a

spin that is due to the motion of its internal parts about its centre. Quantum spin is much more abstract than that and cannot be visualized at all.

All elementary particles that make up matter, such as electrons, protons, and neutrons – and their constituent quarks – are said to a have a spin of one-half (measured in multiples of the value of Planck's constant) and belong to a class of particles known as fermions. Photons belong to the other class of particles, known as bosons, which have the property of having a spin equal to a whole number of Planck constants. There turns out to be a fundamental and crucial difference between fermions and bosons, as is explained in the box on the exclusion principle.

By passing an electric current through hydrogen gas inside a discharge tube, the gas is heated up and radiates light. This is then broken up into its component colours by the prism. But unlike sunlight, which shows a smooth rainbow spectrum in which the colours blend into each other, we now see a series of separate coloured bands (called a line spectrum): red, blue, and many in the violet region, getting closer together until they fade away beyond the visible.

level. Quantum mechanics predicts the frequency and intensity of this light that is re-emitted by excited atomic electrons. Each type of atom has a unique level scheme of allowed electron energies and the patterns of light given off by atoms as their electrons drop down to lower energies are called line spectra. This is how astronomers can know what elements exist in stars and distant galaxies by studying only the nature of the light they see. Today 'spectroscopy' has a wide range of applications.

Into the nucleus

A few years after his discovery of the atomic nucleus, Rutherford carried out an experiment in which he bombarded nitrogen atoms

with alpha particles. He found that the process was causing hydrogen nuclei to be knocked out from the nuclei of nitrogen. These carried the smallest unit of positive electric charge, equal and opposite to that of the electron. He called these hydrogen nuclei protons and revived an old idea due to an English chemist. William Prout had suggested, in 1815, that atoms of all elements were multiples of the lightest one, hydrogen. Rutherford proposed that this might not have been so far from the truth after all. Perhaps it was the atomic nuclei instead that had this property; nuclei belonging to different elements were simply multiples of hydrogen nuclei: single protons. Thus helium nuclei would contain two protons, lithium three, and so on up the periodic table.

But this could not be the whole story. Since protons carried the same amount of electric charge as electrons there had to be as many protons in the nuclei as there were electrons outside them (to ensure atoms were electrically neutral). But nuclei seemed to be much heavier than the sum of protons they contained. If you didn't know the answer to this puzzle what might you suggest?

During the 1920s scientists proposed what they regarded to be a clever trick. Maybe nuclei were made up of protons and electrons – as distinct from the orbiting electrons of course. The total number of protons would be whatever was required to give the right mass (since electrons are so light we can forget about their contribution) but there would be just enough electrons to cancel out the positive charge of the excess protons.

Unfortunately, this idea turned out to be wrong. Yet again, quantum mechanics was able to point in the right direction. Heisenberg's uncertainty principle tells us that trapping an electron within the tiny confines of a nucleus would imply knowing its position to high accuracy. This corresponds to its momentum being too erratic for the attractive force of the protons to keep it there for long. Electrons simply could not be imprisoned inside nuclei.

The uncertainty principle is not showing favouritism towards protons by allowing them the honour of living inside atomic nuclei. It is just that their much greater mass means that they move around

Pauli's exclusion principle

One of the greatest chemists of all time, Dmitri Mendeleev, a Siberian and the youngest of a family with between fourteen and seventeen children (records are unclear), invented the now-familiar periodic table of the elements in the late 1860s. In it, he was able to group all elements of similar chemical properties into families. However, the origin of these groupings remained a mystery for over half a century until the Austrian prodigy Wolfgang Pauli came up with his famous 'exclusion principle'.

Pauli explained that the reason elements have different chemical properties was due to the way their electrons occupied different quantum orbits, or shells. Each electron is described by the quantum numbers assigned to its wavefunction. These define the values of its quantized energy, orbital angular momentum, and spin. Pauli pointed out that no two electrons in the same atom have the same quantum numbers. Once a particular quantum state is 'occupied', the other electrons must find somewhere else to 'sit'.

The exclusion principle also explains why electrons cannot be squashed down into the nucleus, and hence how matter exists in the form that it does.

In this way, a problem highlighted by Bohr could be explained. Clearly, if all the electrons in an atom dropped down to the lowest energy then all elements would have the same chemical properties. An element's properties are not governed by the total number of electrons in its atoms but by how the outermost ones are arranged. Electrons fill up successive 'shells', with each shell accommodating a number of electrons according to certain rules involving the quantum numbers. Once a shell is full, the next electron has to occupy

the next one up in energy. The outermost (valence) electrons then tell us how the atoms bond together to make the seemingly infinite variety of chemical compounds in nature, as well as explaining many of their physical properties – such as why certain materials conduct heat and electricity better than others.

Particles such as electrons, protons, and neutrons, that are collectively referred to as fermions, obey Pauli's exclusion principle. The other class of particles, the bosons, do not. Photons for instance have no problem with all occupying the same quantum state, and indeed prefer to. A popular analogy is to compare the orderly behaviour of fermions with the audience at a classical music concert all seated in regular rows, while bosons can be likened to the crowd at a pop concert who all want to rush up as close as possible to the stage.

more slowly than electrons with the same momentum, and so the fluctuations in their momentum do not give rise to large enough fluctuations in their motion.[5]

The problem was solved when, in 1932, James Chadwick discovered an entirely new particle that was dubbed the neutron. It had about the same mass as the proton but no electric charge. Heisenberg then proposed that atomic nuclei were probably made up entirely of protons and neutrons, and everything suddenly made sense.

What next? Clearly there must be a new kind of attractive force that will glue protons and neutrons together within the nucleus. It had to be unlike anything else encountered in physics before. The mutual electrical repulsion felt between the protons, and the attraction of opposites between protons and electrons, is a form of the well-known electromagnetic force.[6] This, together with the gravitational force, is responsible either directly or indirectly for almost all natural phenomena. All materials are held together via the electromagnetic forces between atoms. On a larger scale, it is the force of gravity that holds our Universe together.

But deep inside the nuclei of atoms we find a new type of force. In 1935 the Japanese physicist Hideki Yukawa came up with a Nobel Prize-winning idea to explain this nuclear glue. To do so, he relied on an idea that was later to become one of the key ingredients in the field of particle physics: that of an exchange particle. To explain what this means, I will need to call upon two now-familiar concepts: Heisenberg's uncertainty principle and Einstein's $E=mc^2$ equation.

Creating particles from thin air

Remember that the uncertainty principle is more generally a statement about our inability to assign precise values to two

[5] Momentum is given by the product of mass and velocity. For a given momentum, if the mass is large then the velocity has to be small.

[6] Magnetism is another form of this force.

complementary quantities simultaneously, such as a particle's position and its momentum. It's a bit like wanting a coin to land both heads and tails at the same time. There are other pairs of quantities that fall into this category, such as a particle's energy and the precise duration that it has that energy. The more accurately we know the energy, the less certain we can be about how long it has this energy. Likewise, the shorter the time interval we consider, the wilder the fluctuations in energy we will get. Now, unlike the position/momentum connection, which led us into deep water with wavefunctions and the nature of what exists when we are not looking, here there is a very definite and beautifully simple consequence: that of particles being created out of absolutely nothing!

In the quantum realm, the time scale of events is very very short indeed. I'll give you one of those mind-blowing examples that scientists are fond of using and which require a quick calculation involving an obscene number of zeros. The number of times that a proton can travel from one side of an atomic nucleus to the other in a single second (moving comfortably within the intra-nuclear speed limit) is many thousands of times more than there have been seconds since the Big Bang.[7] This short time scale allows particles such as protons and neutrons to utilize the uncertainty principle in a neat yet crucial way. They can borrow energy from literally nowhere for a very short duration, provided the energy gets paid back before the uncertainty principle is violated. The shorter the time the energy is needed for, the more can be borrowed.

Next comes phase II: Einstein's equation tells us that mass and energy are interchangeable, so the borrowed energy can be used to create a particle of a certain mass. Yukawa proposed the creation of

[7] The Big Bang was something like 400000000000000000 (or 4×10^{17}) seconds ago, while a proton could cross a nucleus 1000000000000000000000 (or 10^{22}) times in a single second. To travel across once takes one tenth of a time interval called a 'zeptosecond' – a thousandth of a billionth of a billionth of a second. I must say that this is a quite excellent, yet remarkably underused, word and suggest we all try to make much more use of it in the future, as in: 'I'll be back in a zeptosecond', or 'it was over in a zeptosecond'.

At the quantum level, even empty space is not truly empty but is seething with activity; virtual particles are constantly popping in and out of existence everywhere. In pair creation, a particle and its antimatter partner are created out of pure energy, from a photon say. In the reverse process, called pair annihilation, the particle and antiparticle collide and mutually destruct, disappearing forever in a flash of light.

such a particle, now called a pion, inside nuclei. This particle, he suggested, was responsible for the attractive force holding protons and neutrons, known collectively as nucleons, together. Yukawa's calculations predicted that a pion is created by one nucleon, which has borrowed enough energy from its surroundings to create it. The pion then jumps across to a nearby nucleon where it vanishes again. During its brief existence, allowed by the uncertainty principle, it is interpreted as being exchanged by the two nucleons, and giving rise to an attractive force that pulls the two together. Such a pion is often referred to as a virtual particle, implying not real, since it can only have a fleeting existence.

In a similar way, we can think of the electromagnetic force between charged particles as the exchange of a virtual photon – as distinct from real photons that can keep their energy as long as they like, provided they do not get absorbed by an atom of course.

Virtual particles that can be created out of pure energy are known as bosons. They are also referred to as force-carrying particles, in the sense that when they are exchanged between two other particles the process leads to a force between them. Bosons obey different quantum rules to proper matter particles, known collectively as fermions, like

electrons, protons, and neutrons, which make up atoms and hence all the matter we see around us. But, thanks to the work of one of the greatest theoretical physicists of all time, a shy Englishman by the name of Paul Dirac, we know that even fermions can be created out of nothing (see box on 'antimatter').

Nuclear forces

It is fair to say that the study of the atomic nucleus has been, and still is, one of the most challenging fields of human endeavour. The reason for this is the complex nature of the forces that act between the nuclear constituents, the protons and neutrons. Of the four known forces of nature, three are important inside the nucleus. You have already met the electromagnetic force, that acts to push protons apart – since like-charges repel – and what is called the 'strong nuclear force' that pulls all nucleons (protons and neutrons) together. There is also a second nuclear force known simply as the weak force, that is responsible for beta decay, which I will return to in a moment.

It is the interplay between the repulsive electromagnetic force and attractive strong nuclear force that gives rise to the stability of nuclei. Due to the way the strengths of these two forces change over distance, their combined effect conspires to form an energy barrier at the nuclear surface called the Coulomb barrier. This is basically a force field that acts to contain the protons within a certain volume.[8]

Similarly, a positively charged particle hitting the nucleus from outside can penetrate it if it has sufficient energy to punch its way through the Coulomb barrier. But there is a more interesting way that such particles can penetrate the barrier, even when they do not have sufficient energy. This is where we meet yet another quantum

[8] Neutrons do not feel the electromagnetic force since they are neutral and thus oblivious to this energy barrier. However, they are still held in place within the nucleus by the strong force.

Antimatter

Along with Heisenberg, Pauli, and several others, Paul Dirac was another of the young 'brat-pack' of geniuses who placed quantum mechanics on a firm mathematical footing. In fact, a recent poll placed Dirac as the second-greatest English physicist of all time after Isaac Newton.

It is worth noting that Dirac was one of the few big names at the famous Solvay conference of 1927 not interested in arguing over the different interpretations of quantum mechanics. He was more concerned with the aesthetic beauty of the mathematical equations than what they meant!

In 1927, Dirac clarified how the two different formulations of the theory, due to Heisenberg and Schrödinger, were mathematically equivalent. He was then the first to combine quantum mechanics with Einstein's special theory of relativity by devising an alternative equation to Schrödinger's that described the behaviour of electrons moving at speeds approaching that of light. However, Dirac's equation made a strange prediction: that a mirror particle of an electron, or its anti-particle, must also exist. It would have the same mass as the electron but opposite charge. Such a particle was dubbed the positron, and was discovered experimentally a few years later. The positron is also referred to as the electron's antimatter partner.

We now know that every elementary particle has its associated antimatter partner. When the two come into contact, they completely annihilate each other in a burst of energy, since all their properties cancel out apart from their mass, which is converted into pure energy. The amount of energy created obeys Einstein's $E=mc^2$ equation.

This process can also happen in reverse, whereby pure energy is converted into matter: a photon, which is after all a lump of light

energy, can transform itself into an electron and positron in a process known as pair creation.

Most interestingly, particle/anti-particle pairs are continuously popping in and out of existence everywhere by borrowing the energy required for their creation from their surroundings in accordance with the energy/time uncertainty relation, and living for a very short time before annihilating again and repaying the borrowed energy, as though they had never existed.

concept that turns out not only to explain alpha particle decay in radioactivity but is also the reason the Sun shines and how you are here today.

The new concept is known as quantum tunnelling (see box page 168). When an alpha particle, which we now know to consist of two protons and two neutrons, is emitted by a nucleus, it must first overcome the Coulomb barrier to escape. But if we were to apply the familiar ideas of Newtonian physics to the atomic nucleus we would find that it would be impossible for this to happen. A group of two protons and two neutrons stuck tightly together would never acquire between them sufficient energy to escape the nucleus.

A simple way of describing how alpha decay takes place is in terms of the uncertainty relation between energy and time, as explained in the box. But if you prefer, and in keeping with the spirit of the discussions in this book, it can also be understood by thinking about the alpha particle's wavefunction.

Let us begin with a radioactive nucleus that we know for sure has not yet emitted an alpha particle. The whereabouts of the alpha is described by a wavefunction that is confined to the nuclear interior, by which I mean there is zero probability at this time of it being found outside the nucleus. But rather than think of the alpha particle as rattling around inside the nucleus like a tiny ball inside a box until it acquires sufficient energy to escape, we say that its wavefunction begins to leak out of the nucleus. After a short time, the probability of detecting the alpha outside is small; most of the wavefunction is still inside, thus giving a higher probability for no decay. But over time, the probability we calculate from the part of the wavefunction that has leaked out of the nucleus grows to be significant.

By solving the Schrödinger equation to give us the wavefunction at different times, we find that the probabilities we get agree exactly with the observed radioactive decay formula. This states that, starting with a large number of radioactive nuclei, we will find that exactly half of them will have decayed after a particular interval known as the 'half-life'. After a further half-life, only a quarter of the original sample will remain, and so on. But the probabilistic nature of the information we obtain from the wavefunction explains why we can never predict exactly when a particular nucleus will decay.

Unlike the strong nuclear force that binds protons and neutrons together in nuclei, the weak nuclear force has a very different job. It is responsible for causing the second type of radioactivity, known as beta decay. Beta particles come in two types: electrons and their antimatter partners, positrons. Unlike alpha particles, which can be regarded as pre-existing within nuclei since they are also made up of nucleons, the electrons and positrons have to be created in special processes.

Certain nuclei do not have the optimum ratio of protons to neutrons to maximize their stability. To redress the balance,[9]

[9] Nuclei of the lightest elements tend to prefer equal numbers of protons and neutrons, whereas heavier nuclei will have more neutrons than protons.

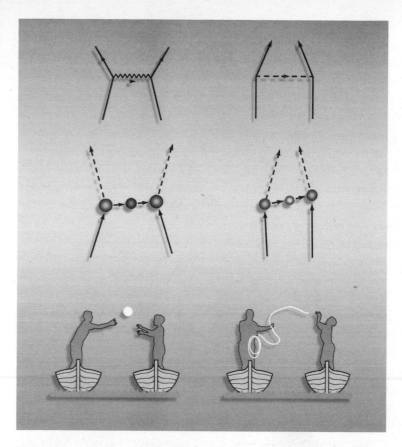

Forces between particles – both attractive and repulsive – can be thought of as being mediated by the exchange of a third particle.

The same process can be viewed in different ways:

Top: Physicists describe particle interactions using 'Feynman diagrams'. The solid, wavy, and dashed lines denote the paths taken by the different particles, and the way the particles' positions change with time can be seen by scanning up the page. Thus the left hand diagram describes how two electrons initially come together, exchange a photon and repel each other. On the right, a proton and neutron stay the same distance apart until they exchange a pion and are attracted together.

Middle: A more schematic diagram of the same processes.

Bottom: The repulsive force between the electrons can be thought of as two people in boats throwing a ball across. First the thrower is pushed back, then the catcher. Attraction would be akin to one person throwing rope across that is caught by the other to pull them together.

Quantum tunnelling

Quantum tunnelling, also called barrier penetration, is another of the bizarre phenomena that only take place in the quantum world. Consider the following example. For a ball to roll up a hill and down the other side, it has to be given enough energy to begin with. As it climbs the slope it will gradually slow down and, unless it has enough steam to reach the top, will stop and roll back again. But if the ball were to behave quantum-mechanically there would always be a certain probability that it would spontaneously disappear from one side of the hill and reappear on the other side. This would happen despite it not having sufficient energy to have reached the top to go over the sensible way.

The standard way of explaining how quantum tunnelling takes place is by appealing to Heisenberg's uncertainty relation between energy and time: provided the energy barrier that a particle needs to tunnel through is not too high or wide, it can borrow sufficient energy from its surroundings to get through. This is allowed as long as it gives back this energy within a time set by the uncertainty relation.

More accurately, we must think of the particle's wavefunction as existing as a superposition of being on both sides of the barrier at once. And it is the wavefunction that is penetrating the barrier. Only when we look do we 'collapse the wavefunction' to find the particle on one side or the other.

Quantum tunnelling plays an important role in many processes. In explaining how alpha radioactivity takes place it became the first successful application of quantum mechanics to nuclear problems. It also forms the basis of many modern electronic devices such as the tunnelling diode.

An everyday example of tunnelling takes place in household aluminium wiring. A thin layer of aluminium oxide will build up on the surface of

In everyday life, if a ball is rolled up a hill without quite enough energy to get it over the top, it will naturally roll back down again. Quantum tunnelling would be like the ball suddenly disappearing half way up the hill and reappearing on the other side. While we never see such magic in the macro world, such a process takes place routinely in the quantum world. Of course, in that case the hill is an energy hill and can be thought of as a kind of force field that the quantum particle should not be able to surmount.

exposed electrical wiring, forming an insulating layer between two wires twisted together to make a connection. According to classical physics, this should stop the flow of current. But the layer tends to be thin enough for electrons to tunnel across easily, maintaining the flow of current.

neutrons or protons may be converted into each other, and an electron or positron is created in the process to make sure that electric charge is conserved. Thus, a nucleus with too great an excess of neutrons will undergo beta decay, in which a neutron will change to a proton plus an electron, which is emitted from the nucleus. An excess of protons prompts one of them to convert into a neutron plus a positron that carries off the proton's charge.

In 1933 Wolfgang Pauli realized that this process must involve the creation of another, then undetected, particle in order to explain why the emitted beta particles did not emerge with the correct energy to balance the books. This new and highly elusive particle, not confirmed experimentally until 1956, is called the neutrino.[10]

Both beta and alpha decay are ways in which nuclei can transform from one species into another. These, together with the processes of fission and fusion, in which nuclei split in half or stick together respectively, form some of the processes that explain how the different elements we see around us today were first formed, including those that make up the atoms in our bodies. The carbon, oxygen, and nitrogen, along with the other elements that go into forming the chemicals necessary for life, would have been synthesized inside stars billions of years ago. Those stars no longer exist since they have exploded as supernovae with much of their contents thrown out into space, ending up forming part of our Solar System. This is the reason for the oft-quoted yet quite true statement that we are all made from star dust.

Many heavier elements are only made when a massive star explodes violently as a supernova. The hotter and the more extreme the conditions in a star, the further the synthesis process is able to go and the heavier the elements that are formed. The interior of a star only becomes hot and dense enough to make the heavier elements

[10] Millions of neutrinos from outer space pass through every square centimetre of your skin each second of your life and carry on straight through your body without 'touching' you. The same thing happens when they travel through any material, so it is not surprising that they are so hard to catch and study.

during these final intense moments of the star's life.[11]

The two lightest elements, hydrogen and helium, were created not inside stars but in the very early Universe just after the Big Bang. Roughly 98 per cent of all the visible matter in the Universe today is composed of these two elements, with the remaining 2 per cent for everything else.

The rich variety of nuclei that can exist, based on the different ways that protons and neutrons can be combined together (and we have so far studied just a few hundred of the seven thousand or so possible species)[12] is due to two reasons. The first is that protons and neutrons obey similar quantum rules to the orbiting electrons that dictate the way they can arrange themselves within nuclei. Just as electrons must be described by wavefunctions whose shapes are determined by the quantum number labels they carry, so too should nucleons be more correctly viewed as spread-out entities distributed within the nucleus according to their quantum numbers.

At least with the electrons we have the luxury of a mental picture of tiny orbiting balls, even if this is not correct. Inside the nucleus space is at a premium and the nucleons are packed together such that the only picture we can conjure up is that of a bag full of balls jostling about for position.

In practice, the picture we have of a nucleon depends on how we try to probe it. If we are interested in describing how high-energy protons or neutrons fired into a nucleus interact with the other nucleons, we discover that it is reasonably accurate to treat them as tiny localized particles. But an outer neutron in a halo nucleus (see box in Chapter 3) has a wavefunction that is smeared over a large volume around the whole nucleus.

The second reason for the complexity of nuclei lies in the nature

[11] For a colourful and lucid description of how the elements are created in the cosmos look no further than *Nucleus: A Trip Into the Heart of Matter*, by Ray Mackintosh et. al. (Canopus Publishing, 2001).

[12] Think of it roughly as a hundred or so elements (differing in their number of protons), most with tens of different isotopes (differing in their neutron number).

of the strong nuclear force, which turns out to have an even more fundamental origin than Yukawa anticipated with his pion exchange picture. By the second half of the 20th century, physicists began to wonder whether something deeper was going on inside nucleons.

Quarks

By the mid-1930s a handful of elementary particles were known to exist. Apart from the protons, neutrons, and electrons that made up the atoms of normal matter, and the photons of electromagnetic radiation, physicists had also discovered positrons and neutrinos. Then, soon after Yukawa came up with his pion theory, a new particle was detected in cosmic rays, which was first mistakenly thought to be Yukawa's pion. In fact it resembled a heavy and unstable electron and is today known as the 'muon'. Muons are formed in the Earth's upper atmosphere when high-energy protons arriving from outer space collide with air molecules, and have a lifetime of just a fraction of a second. Pions were later discovered in experiments a few years later.

Particle accelerators (or atom smashers as they were known at first) were soon built to probe deeper into the structure of the quantum world. The idea was simple: rather than use light to look inside subatomic structures, physicists followed Rutherford's lead with his alpha particles. But to probe smaller length scales they needed to use more energetic particles. Essentially, they used the wave-like properties of matter particles instead of light waves. The higher the energy of the particle beam, the shorter its de Broglie wavelength and the smaller the length scales it would resolve. In addition, the more energy that could be released from a tiny volume, from increasingly violent particle collisions, the likelier it became for ever-more-exotic particles to be created out of this energy.

By the second half of the 20th century, so many new elementary particles had been discovered that physicists began to question whether they were truly elementary at all. Just as the atoms of the

92 different kinds of elements were found to be made up of just three particles: protons, neutrons, and electrons, so perhaps were all these particles composed of just a few even more fundamental constituents.

In classifying the particles, one family in particular seemed to contain too many varieties. Hadrons are the particles that feel the strong nuclear force, and they are classified into two groups. The first, called baryons, includes the proton and neutron. To these were soon added a whole host of new baryon particles, such as the 'Lambda', 'Sigma', 'Xi' and 'Omega'. The second group, known as mesons, contains the pion along with a number of other more massive particles, such as the 'eta' and the 'kaon'.

In an attempt to restore simplicity and economy, two theorists, Murray Gell-Mann and George Zweig, proposed that maybe all hadrons (baryons and mesons) have internal structure. They showed how all the different varieties could be understood if they were composed of more elementary particles called 'quarks'.

Just a few years later, this hypothesis was confirmed at the Stanford Linear Accelerator in California. In an experiment that was remarkably similar in approach to Rutherford's famous alpha-particle scattering set-up that confirmed the internal structure of atoms, high-energy electrons were scattered from protons and neutrons. This time, the directions in which the electrons bounced back revealed that hidden inside each nucleon were three tiny lumps of matter. The quark idea had been vindicated.

Initially, it had been thought that there were just three types (known as flavours) of quark. We now know there to be six in total, each with a different mass. Protons and neutrons are made up of just two types, with rather unimaginative names: a proton contains two 'up' quarks and one 'down' quark, while a neutron contains two 'down's and an 'up'.

It turns out that the unit of charge carried by a proton or electron is not the smallest bit of electricity there is. Three of the quarks are negative and have one-third of the charge of the electron, and the other three are positive with two-thirds of the proton's charge. In

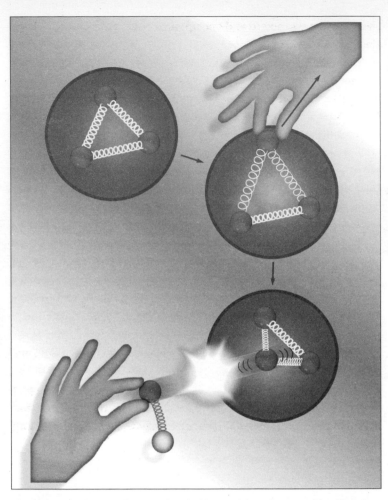

We can never pull out single quarks from inside a particle such as a nucleon. Even if we provide enough energy to overcome the binding between the quarks, all we succeed in doing is supplying energy to create a new quark/antiquark pair through the process of 'pair creation' (see figure on page 162). The new quark replaces the removed one inside the nucleon while the antiquark joins the removed quark to form a meson.

this way, two up quarks, each with a positive charge equal to two-thirds that of an electron, and one down quark with a negative charge one-third of the electron's, combine to make up the proton's

charge, while two downs and one up cancel each other out inside the electrically neutral neutron.

The other four flavours are known, for no good reason, as 'strange', 'charm', 'top', and 'bottom'. Personally, I prefer Terry Pratchett's description in his Discworld stories of the four flavours of magic: 'up', 'down', 'sideways', and 'peppermint'!

In addition to electric charge, quarks must also carry another kind of property known as colour charge. This is required in order to explain why quarks only ever group in threes to make up nucleons and all other baryons, but in quark/antiquark pairs to form the pion and its meson relatives. I will say more about this in Chapter 8.

Today there are known to be just two species of elementary matter particles: quarks and leptons. The latter is the name for all particles that do not feel the strong nuclear force – that is all those that do not carry colour charge – in other words all elementary matter particles that are not quarks! Leptons include the electron and its two heavier relatives, the muon and the 'tau', as well as three types of neutrino.

At least it is nice to know that the very first elementary particle to be discovered, over one hundred years ago, is still elementary. Let's hear it for the electron.

So how sure are we that electrons and quarks are the most fundamental building blocks of matter? Maybe in time we will find that they too have internal structure. Maybe there is something even more basic and fundamental.

Fermions are comprised of three generations of particle families – quarks down the left hand column and leptons on the right. All matter in the Universe that is composed of atoms is made from first generation particles only (shown on the surface). This family consists of the up and down quarks, which make up the nucleons inside atomic nuclei, the electron and its neutrino. The second and third generation particles are much heavier and only live very briefly. These can be produced inside particle accelerators.

Elementary Constituents

Frank Close Professor of Physics, University of Oxford

At various stages in history the candidates for the fundamental building blocks have differed. A century ago, the atomic elements were believed to be fundamental; by the 1930s it was electrons, protons, and neutrons. Today, the electron is still on our list but protons and neutrons are known to be made from yet smaller particles – the quarks. An obvious question, with history as our guide, is whether the electron and quarks are indeed fundamental, or made of yet smaller pieces like Russian dolls. The honest answer is: we don't know! All we can say is that with the best experiments we are able to do today, there is no hint of deeper structure. There are also hints that there is something special about this layer of the 'cosmic onion'.

How did we come by this route? There are two experimental techniques; one is by scattering, the other is spectroscopy.

If a supposed fundamental layer is actually made of deeper constituents, quantum mechanics restricts the ways that those constituents can arrange themselves. One of these configurations will have the least amount of energy: we call this the ground state. One or more constituents may be in a higher state of energy, causing the system as a whole to have higher energy. A constituent may emit a photon of light, losing energy in the process; conversely, absorbing a photon of suitable energy can raise the system from its ground state to a state of higher energy. From the spectrum of photon energies, one can deduce the pattern of energy levels of the (composite) system.

The resulting energy levels of a molecule (due to its atoms vibrating about one another); an atom (due to its electrons); a nucleus (due to the vibration and orbiting of its protons and neutrons) and even of the proton or neutron itself (due to the motions of their constituent quarks) all look qualitatively very similar. However, quantitatively, they differ.

The units of energy on the quantum scale are 'electron Volts': 1 eV = 1.6 x 10⁻¹⁹ Joules. To give you an idea of scale, it typically takes several eV of energy to pull an electron from an atom. The energy scale for molecular excitations is in milli-electron Volts (denoted meV); atomic nuclei are excited on the MeV (millions of electron Volts) scale and the protons or neutrons on hundreds of MeV. These reflect the ever smaller distance scales and forces at work as one goes from relatively large molecules down to compact protons. This is the first clue to a deeper structure. Then, scattering directly from those constituents (such as Rutherford's experiments on the atomic nucleus, or high-energy scattering of beams of electrons on the quarks) reveals their inner constituents.

Suppose we list the varieties of quark or the electron, muon and tau analogously. The lightest ones have masses of the order of MeV . The tau, charm quark, and bottom quark are all on the scale of GeV (billions of eV) while the top quark is hundreds of GeV. Could this be a sign of new spectroscopy due to 'subquarks' and 'subleptons'?

Following the old familiar routine does not work, however. There is no sign of electromagnetic transitions (the emission or absorption of photons) between the 'heavy' and 'light' leptons for example, contrary to what would happen if they were simply excitations of one another. The same holds true for the quarks, though the evidence is less direct. Second, all these particles have the same amount of spin (1/2 units of Planck's constant) and one would expect a whole range of such spins for an excitation spectroscopy. Also, there are indirect hints that there are at most three 'generations' whereas a simple excitation spectrum

would have an entire range of states.

Finally, their size (if any) is less than 10^{-18} m and at such minuscule dimensions one would expect all of their masses to be on the scale of many GeV, rather than MeVs (in the case of the up and down quarks and the electron).

Either these particles are truly fundamental, or there are dynamics at work beyond conventional quantum mechanics. Either one of these options is exciting. There does appear to be something novel about the present family of 'fundamental' particles that appear in the Standard Model.

Chapter 8
The Search for the Ultimate Theory

In order to understand the structure of our universe at its deepest level, physicists look to answer every question and solve every mystery. We never stop asking 'why?' –

Why did this happen?

Because of such and such an effect.

What was the cause of that effect?

This body interacting with that one

Why did they interact?

Because they feel such and such a force

What is the origin of the force?

And so on. Unlike children, however, we cannot be pacified by an exasperated parent's conversation-ending remark of 'because God made it that way!' Of course, many scientists have a religious faith, but this hardly ever conflicts with their quest to seek answers to the most fundamental questions in their field of research.

Theoretical physicists, however, are driven by more than a thirst for knowledge through delving deeper and deeper into the complex workings of nature. They also look for patterns and symmetries in nature that manifest themselves in the simplicity and beauty of

mathematical equations. Some of the greatest minds have even gone so far as to reject a theory because the mathematics was too cumbersome or ugly! They would say something like: there must something wrong here; surely nature cannot really be this awkward. Now if you are not a mathematician or a physicist this might seem like irrational grounds for discarding a theory, but it does seem to work. The quest for the ultimate truths is always a quest for beauty and simplicity. It does appear that the multitude of phenomena we observe around us, whether on Earth or deduced from the light of a distant star, are ultimately explained by a remarkably small number of fundamental theories. The whole of classical mechanics is explainable by Newton's laws of motion and forces, and Einstein's theories of relativity improved on this; both electricity and magnetism turn out to be two manifestations of the same electromagnetic force; and of course the behaviour of all subatomic particles is described by quantum mechanics.

So 20th-century physicists needed to do more than simply find and classify all fundamental particles. They needed to understand the way these particles interacted with each other and the origin of the forces between them. And if there were different types of forces, did these have a common origin? The quantum mechanics of the 1920s was really just the first step along this path. The story of the progress in atomic, nuclear, and particle physics I told in the last chapter is thus incomplete. For, in their quest into the structure of matter to find its fundamental building blocks, physicists have also been searching for the simplicity and symmetry in their theories. The Holy Grail of physics is to find the ultimate theory of everything, an all-powerful theory from which all naturally occurring phenomena in the Universe could be deduced and explained.[1]

In this chapter we will trace the progress that has been made in

[1] A note of caution about reductionism: most phenomena in nature are complex and many of their properties only emerge on the macro scale. Thus studying a single H_2O molecule gives no hint of the 'wetness' of water.

this direction, and assess whether we are finally closing in on such a theory.

The quantum theory of light

So far, when I have discussed the issue of wave-particle duality it has been in the context of quantum objects such as electrons behaving as waves when we are not observing them and particles when we do. But the wave-like aspect is that of the wavefunction, and I do not want to repeat the controversial arguments surrounding the physical meaning of the wavefunction again. When it comes to light (and indeed all other forms of electromagnetic radiation), the wave aspect is quite real. We seem to have the choice of regarding light either as a physical wave or as physical particles, depending on how we look at it and what sort of phenomenon we study.

In fact, the advent of quantum mechanics did not persuade physicists to abandon their classical wave theory of light. This theory, due to Scotsman James Clerk Maxwell in the second half of the 19th century, is encompassed in a set of equations that bear his name. Maxwell showed that light is made up of a combination of electric and magnetic fields oscillating at right angles to each other travelling at a speed of 300,000 kilometres (186,000 miles) per second.

An important property of Maxwell's equations was that they were consistent with special relativity. Unlike Newton's equations of motion, which had to be modified by relativity in order to be correct at speeds approaching that of light, Maxwell's equations were already correct. Of course, a theory that describes something moving along at the speed of light (namely light itself) that didn't agree with special relativity would have posed something of a dilemma.

In contrast, quantum mechanics, as proposed by Heisenberg and Schrödinger, was not consistent with special relativity. It could only describe the behaviour of quantum entities like electrons that moved at speeds well below that of light. So just as the motion of a classical object such as a planet, or a football, can be accurately described by

Newton's equations, but needs modifying if it were to approach light speed, so too is Schrödinger's equation only applicable to slowly moving quantum objects. At relativistic speeds, special relativity shows how quantities such as the mass, momentum, or energy of a body begin to change. Therefore, at the very least, the values of these quantities in the Schrödinger equation had to be replaced by their relativistic versions.

I'll give you a quick example of what this entails. The mass of an object implies the amount of 'stuff' it contains. In fact, in everyday language mass is taken to mean the same thing as weight.[2] We thus regard mass as a constant quantity that does not change just because a body is moving. However, special relativity has taught us that as the body approaches the speed of light its mass starts to grow until, at light speed, it would have an infinite mass – which is why nothing that has mass when at rest can travel at the speed of light. Just one year after Schrödinger published his original equation, its reformulation with these changes was made independently by Oskar Klein and Walter Gordon, as well as Schrödinger himself. But the new equation had a rather serious problem: the quantum probabilities it predicted from its wavefunction could be negative! What on Earth could it mean to say that an electron had a minus twenty per cent probability of being somewhere?

In 1928, Paul Dirac published a paper with the title *The quantum theory of the electron* in which he proposed an alternative equation to Schrödinger's which was not only 'fully relativistic' but also took into account the electron's spin in a natural way (something that was important at the time if theory was to explain new experimental results). It was this equation that led Dirac to the theoretical prediction of anti-particles[3] and the idea of electron-positron creation and annihilation.

A year earlier, in 1927, Dirac had also published the first

[2] The difference is that, out in space, you would have no weight as there is no force of gravity, but your mass would be the same.

[3] First proposed in a letter Dirac wrote to Bohr in 1929.

pioneering paper combining quantum mechanics with Maxwell's theory of light to give the first quantum theory of the photon. What he did was 'quantize' the electromagnetic field.

He then worked out how to combine these two theories, one describing the electron and the other the photon. What emerged was a theory of quantum electrodynamics. It was the first example of what is known as a quantum field theory and explained how electrons emit and absorb photons, and how two electrons will repel each other by exchanging a photon.

After a promising start, quantum field theory had a rough time during the 1930s and 40s when it was plagued by mathematical difficulties. You see, unlike the old quantum mechanics, it allowed for virtual particles to be created and destroyed all the time – remember this is what you get when you combine Heisenberg's uncertainty principle with Einstein's $E=mc^2$. This meant that certain calculations using the theory led to infinite answers. I can explain how this comes about in a rough hand-wavy sort of way. The basic idea of a quantum field theory is that something like an electric field can be thought of as many virtual photons popping in and out of existence all the time. So an electron swapping a single photon with another electron is only the simplest process that can happen. If we start probing down to smaller distances we find more and more going on. For instance, that virtual photon might, while en route between the electrons, spontaneously become a virtual electron-positron pair that quickly annihilates to give back the original photon before it reaches its destination. But during the short lifetime of the virtual electron and positron, they should also be able to exchange another virtual photon, and this might create its own electron and positron, and so on. Now you would hope that the possibility of such ever-more complex activity could be ignored in the calculation, or at the very least become progressively less important, but it doesn't. It leads to infinite answers in the calculations.

This problem had actually been around since long before quantum mechanics. In the 19th century, physicists worried about the following situation. An electric charge will generate around it

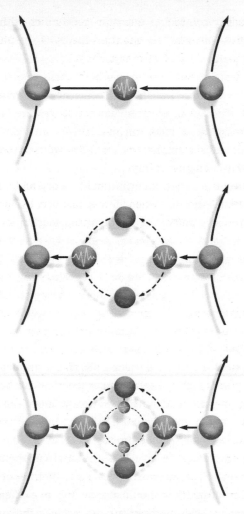

Quantum field theory describes the way two electrons interact by taking into account a series of evermore complex, but less likely, processes that can take place. The 'lowest order' process is that they exchange a single virtual photon (top). Higher order terms that must also be considered involve one of the electrons emitting a photon that creates, whilst in transit, an electron-positron pair. These quickly annihilate to produce a photon again that is absorbed by the second electron (middle). There is even a small probability (bottom) that the electron-positron pair created by the exchanged photon can themselves exchange a virtual photon, which also creates a further electron-positron pair, and so on.

an electric field, but how do we work out the effect this field has on the charge that generated it in the first place? This is only a problem at the place where the charge is situated, but here we would get an infinite answer. This is because we would need to divide a certain quantity by the distance between the point we are interested in and the position of the charge. In this case, that distance is zero. And dividing anything by zero gives us infinity.

These infinity problems that plagued quantum field theory were finally sorted out when, in 1949, three physicists, Richard Feynman, Julian Schwinger and Sin-Itiro Tomonaga, separately figured out a way of finessing the infinities with a mathematical trick called 'renormalization'. What emerged was a theory that is to this day regarded as the most accurate in the whole of science. It is still called quantum electrodynamics – QED for short – but this is what many physicists mean today when they use the term. But we must not forget that QED was first suggested by Dirac twenty years earlier.

QED's agreement with experimental measurements is to one part in one hundred million. But don't go thinking that this is just a neat description of the way electrically charged particles feel each other's presence by exchanging photons. This theory of the nature of the interaction of light with matter is the most fundamental and important in the whole of science. All mechanical, electrical, and chemical laws and phenomena are ultimately derived from it. Apart from the force of gravity, and the forces acting inside the atomic nucleus, all processes in nature are ultimately explained by QED: how hydrogen and oxygen atoms join up to make a water molecule, the nature of sunlight, how the image of this page appears on my laptop screen as I type, and how the electrical signals from my brain are converted to mechanical reflexes that control the motion of my fingers as they tap out the letters on the keyboard.

In this way we see that QED underlies the whole of chemistry (and hence biology), since it fundamentally boils down to the way atoms interact via their electrons – and this is through the electromagnetic force, which is no more than the exchange of photons.

Some physicists, including Dirac himself, didn't like the way

QED's renormalization trick took care of the infinities. It seemed as though, mathematically speaking, the awkward bits were being swept under the carpet. So while the resulting theory worked beautifully, purists such as Dirac always felt this should not be necessary and held out for something more fundamental.

The major goal of researchers in fundamental physics for the past half a century has been something far grander. For all its success, QED is a theory describing just one of the four forces of nature. Is it possible to describe the other three forces (gravity and the two nuclear forces) using a quantum field theory – by which I mean using the concept of the exchange of quantum particles? Better still, is there a single quantum field theory that could do the whole job?

Gauge theories and symmetries

When we use the word 'symmetrical' in everyday language we usually mean something rather specific: that an object or shape looks the same as its mirror image, or when viewed from different angles. But in mathematics, the idea of symmetry has a much more powerful meaning, and has helped physicists in their quest to unify the forces in a quantum field theory.

To give it a more general definition, a symmetry exists if some property remains the same when some other quantity is changed. So a sphere looks the same whichever angle it is viewed from, and the age difference between two people remains the same with the passage of time. These are both forms of symmetry. Physicists speak of a 'global' symmetry when certain laws of physics remain the same when a particular change, or 'transformation', is applied equally everywhere. Certain physical theories have an even neater property. For instance Maxwell's equations of the classical theory of electromagnetism remain the same even when certain transformations are applied 'locally' (that is differing from one place to another). This has to do with the fact that electric and magnetic fields are in some sense equivalent to each other.

To see this, let us represent the electrical potential energy felt by an electron as a hilly landscape with valleys representing attraction, since the electron will roll into them, and hills as repulsion, since if placed on one the electron will roll down and away from it. If the shape of the landscape is changed in one location, say by pushing a valley upwards into a hill, then gauge symmetry requires the electron to behave in exactly the same way as it would have before the change, namely roll towards the hill. But to do this the electron requires a compensating change in the magnetic potential energy. The theory of electromagnetism is said to be a gauge theory with local symmetry.

It turns out that QED has this property. In fact, it was discovered that a quantum field theory for any of the four forces of nature would involve such gauge symmetry. This gave physicists hope that maybe the forces were somehow related to each other.

I know this all sounds rather technical, but I mention it for a reason. What is important in this discussion is the idea of 'breaking' a symmetry. A blank sheet of paper is symmetric under certain rotations; it looks the same on either side or upside down. But as soon as you start to write on it, this symmetry is lost, or broken.

During the 1960s, physicists had used this type of symmetry argument to extend QED to incorporate the weak nuclear force – that is responsible for nuclear beta decay – as well as the electromagnetic force. It was found that under certain conditions the weak force could also be thought of as exchanging virtual particles like the photons. And if certain special symmetries were broken then the old renormalization trick could still be used to make sense of the quantum field theory of the weak force. By the late 1960s, Steven Weinberg, Abdus Salam, and Sheldon Glashow had developed an extended field theory that unified the electromagnetic and weak forces. It is known as the 'electroweak theory'.

They explained that above a temperature of about a million billion degrees, such as would have been the conditions in the very early Universe, the electromagnetic and weak forces would become one and the same force. But as the Universe cooled and expanded a certain symmetry was broken, and two very different forces would

have crystallized. Today, we understand the weak force to be due to the exchange of particles known just by the letters 'W' and 'Z'. Well, they are more accurately called 'weak vector bosons', but W and Z bosons is easier.

A colourful force

As soon as the idea of gauge symmetry was found to apply to quantum field theories, progress was also fairly rapid in understanding the strong nuclear force in this way. Of course, Yukawa had paved the way many years earlier when he proposed his theory of the pion, the particle that can be thought of as being exchanged between nucleons inside the nucleus. But once it was discovered that the nucleons were themselves made up of quarks it was realized that the exchange force had also to be operating down at a deeper level. The field theory of the strong nuclear force that was developed is known as 'quantum chromodynamics', or QCD.

Let me first ask you to think a little about what electric charge really means. At a fundamental level we can do no better than to say that it is a property of certain elementary particles that comes in two types, which we call positive and negative. Particles with opposite types attract each other and like-charged particles repel each other. Alternatively, we could have called the two charges sweet and savoury. A particle with sweet charge will be attracted to another with savoury charge. Anyway, you get the idea; positivity or negativity of electric charge is a made-up convention.

In the same way, the strong nuclear force had to have a convention for naming the property of the particles that felt its influence. They were said to have 'colour' charge. In order to explain the quark model, namely that each nucleon had to contain three quarks, there had to be three types of this strong force charge. The reason the analogy with colour was chosen was because of the connection with the way different colours of light combine. Indeed, the origin of the name of this theory: 'chromo' is derived from the Greek word *chroma* meaning

'colour'. The three colour charge types were thus red, blue, and green. A red, a blue, and a green quark could combine to produce something that was colourless. The rule was that quarks could not exist by themselves because they had colour; only colourless combinations were allowed. This is similar to the much earlier arguments about the nature of atoms, which had to contain equal amounts of positive and negative charge in order to be electrically neutral. Although in that case, nature at least allows for atoms to lose or gain electrons and thus exist as positive or negative ions.

While isolated quarks cannot exist, there is considerable current research going into producing what is known as a quark-gluon plasma, or deconfined quarks and exchange particles called gluons. This forms when two heavy nuclei are smashed together at very high energies. For a fraction of a second the boundaries between the protons and neutrons inside the two nuclei will melt away and we are left with a soup of free quarks and gluons that quickly 'freeze out' into various hadrons. The conditions of very high temperature and density that are necessary to create this plasma are thought to have existed just after the Big Bang.

It is worth noting that the colour-combining tricks of quarks inside hadrons were thought up before quantum chromodynamics came along. A big difference between QED and QCD is that in the former, there is just one type of force carrier: the photon. In QCD, there are eight different types of colour force exchange particles, gluons, to account for the different ways that the coloured quarks can interact. In the last chapter, I discussed how Yukawa had proposed that the pion was the fundamental strong force-carrying particle when it is exchanged between two nucleons. We can see that at a deeper level it is the gluons that are the true strong force carriers.

So, just as the electroweak theory describes particles interacting by exchanging the force-carrying photons or W and Z bosons, so QCD is the quantum field theory of quarks exchanging gluons.

There is, however, still some work to be done in this ambitious programme of unifying the forces. While physicists successfully combined the electromagnetic and weak forces into one theory, they

have still not been properly able to merge the electroweak theory with QCD despite both being quantum field theories. Put it this way – there is a scheme for combining them, but it has yet to be verified experimentally. Until such a time, particle physicists refer to the framework that loosely incorporates both the electroweak theory and QCD as the 'Standard Model'. It works very well, but no one believes it to be the final word on the matter.

Having a GUT feeling

One way of reaching the limit where the different forces become unified is to probe shorter and shorter length scales. QED explains how an electron is always surrounded by a mist of virtual photons, as well as virtual electron-positron pairs popping in and out of existence all the time. All this activity tends to mask the electron's electric charge and ends up producing just the charge we actually see. This is another way of describing how renormalization copes with infinities. Here the infinity is the electron's charge itself, but it only starts to grow big when we zoom in through the surrounding veil of virtual particles.

As we get closer and closer to the source of the electromagnetic force we find that the strength of the force grows. For the two nuclear forces – which are much stronger than the electromagnetic one within the range that they operate (inside nuclei) – the opposite happens: at shorter and shorter length scales, these two forces become weaker. When we reach a distance scale (10^{-28} of a millimetre), that is as small compared to the size of a proton as a proton is to us, we find that all three forces converge on the same strength. It is here that they can be regarded as a single force and a certain symmetry is restored.

A theory that would unify these three forces is known as a Grand Unified Theory (or GUT for short). Physicists have been trying to find a GUT, with everything slotting into place, for a while now. It could then replace the less satisfying Standard Model, the loose alliance between the electroweak theory and QCD. A particularly awkward

problem that was solved back in the 1970s was to do with the length scale when the strengths of all three forces converge. At the precise point in our magnification when the electromagnetic and weak forces converge and the electroweak force kicks in, the strong force is still too strong, and the symmetry remains broken. For true symmetry, the three needed to converge simultaneously.

Then a new kind of symmetry was discovered which was even more powerful than the one that had been needed to unify the electromagnetic and weak forces. It is known as 'supersymmetry' and is a mathematical way of fixing this problem. Basically, it reveals the symmetry, or connection, between electrons, neutrinos, photons, and W and Z bosons on the one hand (the particles described by the electroweak theory), and the quarks and gluons on the other hand (the particles of QCD). Its main prediction is that each of the known particles has a 'supersymmetric' partner with the opposite character. Thus the electron would have associated with it the selectron (a boson) and the photon would have the photino (a fermion). Supersymmetry also predicts that a proton could decay into a positron plus a pion. If this process could be detected in nature, it would be strong evidence in support of a grand unified theory. So far, proton decay has not been seen, but it would be so rare that there is always the chance it may have been missed.

We simply don't know yet whether nature also behaves in a supersymmetric way. But maybe supersymmetry has a more fundamental role to play, in an endeavour that makes grand unified theories seem parochial in comparison.

What about gravity?

Aren't we forgetting something? It seems a little mean, if not downright rude, to talk about unified theories being grand if they only aim to include three of the four forces of nature. I have so far omitted to say anything about how the force of gravity fits into all this. It is not as though people haven't tried – Einstein spent the last

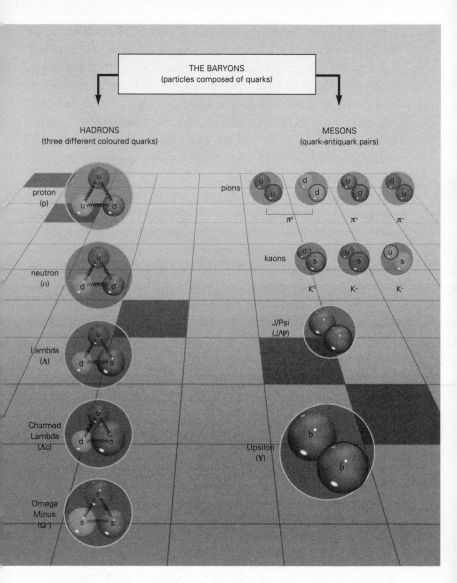

A small selection of the baryons. These are the particles that are composed of quarks and feel the strong nuclear force through the exchange of gluons. Hadrons are composed of three different-coloured quarks held together by the gluons. Mesons consists of quark-antiquark pairs.

All elementary particles can be grouped into two categories: the matter particles (fermions) and the force particles (bosons).

thirty years of his life trying to find a field theory that would unify electromagnetism with gravity, to no avail.

In a sense you should not feel pity for the gravitational force. After all, it is described by a theory that some regard as more beautiful, more powerful, even more fundamental, than any quantum field theory: general relativity. (Cue trumpets, tingle down back of spine etc.)

In his special theory of relativity, Einstein showed that there was no such thing as absolute space and time, since no two observers could agree exactly on distances and time intervals. Only by combining space and time into four-dimensional spacetime can we really make sense of things. In 1915, Einstein completed his greatest contribution to science. General relativity was an extension of his special theory to include the force of gravity. However, the description of this force could not be further from the picture of exchange particles afforded us by quantum field theory to describe the other three forces. Instead, Einstein described gravity in terms of pure geometry. Everything in the Universe is trying to pull everything around it closer. But the attractive force of gravity in general relativity is due to the curvature of spacetime itself. The bigger the mass of the body, the more it warps the space and time around it.

General relativity has been shown experimentally to be exceptionally accurate. It is our best current theory of the nature of space and time themselves; pretty deep stuff!

Physicists now realize that to unify the four forces of nature, they must find a way of combining general relativity with quantum field theory (which is just a fancy way of saying quantum mechanics).[5] The problem is that these two theories have hardly anything in common apart from the fact that they both approximate to Newtonian physics on an everyday scale. They depart from this limit at the extremes of very small objects and distances (quantum mechanics) and very large

[5] Don't confuse the combination of quantum mechanics with general relativity with Dirac's achievement in 1928 of uniting quantum mechanics with special relativity. That was a piece of cake in comparison. After all, special relativity is a theory without forces.

objects and distances (general relativity). But with this departure they take on very different mathematical structures, making them incompatible.

Nevertheless, by the late 20th century the quest was truly on for the ultimate theory, a theory of quantum gravity.

The lesson of Planck

The quest for a theory of quantum gravity, which would unify all four forces, is today an exciting field of theoretical research. I will outline in the remainder of this chapter some of the basic ideas involved.

Workers in the field have divided up loosely into two camps. One camp claims that quantum mechanics contains the more fundamental concepts and that we should start from there and work our way up to incorporate general relativity. The other camp disagrees, and prefers to start from general relativity with its fundamental notions of space and time, and tries to quantize it. Of course there are other physicists who believe that neither will survive fully intact and that both will need major surgery before they can be grafted together into quantum gravity. A yet smaller minority of deep thinkers even proposes that the correct route is to ditch both quantum mechanics and general relativity and start from scratch. But since each theory works so well in its own domain it is hard to believe that they don't both contain fundamental truths about nature.

There is at least one issue that all quantum gravity workers now agree on: we must follow the lesson of Planck one hundred years ago. He started the quantum ball rolling by proposing that energy cannot be indefinitely subdivided but is composed of irreducible lumps, or energy quanta, just as matter is ultimately made up of fundamental building blocks. Quantum gravity workers now realize that space and time themselves must also be ultimately composed of irreducible lumps. And, in honour of the founding father of this idea, the length and time scale where this quantization must

take place is known as the Planck scale. It is at this scale – which can be thought of as a distance scale or an energy/temperature scale – that all the forces are unified.

So what is the size of the tiniest unit of space, a quantum of volume? Let us get this into perspective: there are more atoms in a single glass of water than there are glasses of water needed to fill all the seas and oceans of the world. So, atoms are tiny. Next, you could pack a thousand billion atomic nuclei within the space of a single atom. So, nuclei are far more minuscule. Right then; deep breath; an atomic nucleus can accommodate as many quantum volumes as there are cubic metres in the Milky Way Galaxy (roughly 10^{62} m^3). And our Galaxy is not exactly cramped: eighty thousand light years across and containing a hundred billion stars. It puts a strain on the brain just thinking about it.

This idea of the tiniest unit of space means that it makes no sense to talk about chopping such a volume in half.

And what about the duration of a quantum unit of time? This is so short that I cannot think of an appropriate analogy to impress you with. Suffice it to say that there are many many more quanta of time in a single second (10^{43} of them) than there have been seconds elapsed since the birth of the Universe (less than 10^{18}).

String theory

The majority of quantum gravity workers belong to the camp that is trying to build on quantum mechanics. They argue that we have come so far, using such ideas as quantum field theory and supersymmetry, in understanding three of the four forces in terms of quantum mechanics that surely gravity can also be tamed. In fact, they are developing a candidate theory, called string theory, that does just that. It describes the force of gravity in terms of an exchange particle called the graviton. But string theory, as its name suggests, differs in a remarkable way from the earlier quantum field theories. It states that all the fundamental particles are in fact tiny vibrating strings.

The different frequencies at which these strings vibrate give rise to the different elementary particles.

The original version of string theory was based on ideas developed in 1968 by Gabriele Veneziano. But his theory suffered from a number of problems such as the prediction of faster-than-light particles, called tachyons, which is something physicists no longer accept as a possibility. Then in the mid-1980s, the first string revolution took place. In a landmark paper by John Schwarz and Michael Green, the idea of supersymmetry was applied to Veneziano's strings, and it cured the problems. The new version became known as superstring theory and was heralded as the ultimate Theory of Everything.

After the initial excitement, however, progress in superstring theory research slowed down. There were two reasons for this. Firstly, the mathematics was so complex that no one knew just what the equations meant. In fact, it has been said that some of the mathematical methods needed to solve the equations properly have not even been invented yet! The second problem was more disappointing. By the early 1990s, five different versions of string theory existed and no one knew which was the correct one. The field stagnated somewhat and the experts began to find it hard to convince bright new students to join them to train for their PhDs in the subject. It seemed to have lost its sex appeal.

Then in 1995, the second string revolution took place. A new, even grander framework was proposed which unified the different superstring theories under one umbrella. It is known as 'M-theory', where the 'M' stands for 'membrane'. Instead of just strings being the fundamental entities of our universe, M-theory predicts that there should also be two-dimensional sheets, or membranes, and even weird three-dimensional blobs. But as yet, no one even knows what the equations of M-theory look like, and many prefer the 'M' to stand for 'mystery'. What is known is that M-theory predicts that we really inhabit a ten-dimensional space (plus the dimension of time to make an 11-dimensional spacetime). However, six or seven dimensions of space are curled up far smaller than anything that

can be accessed with today's particle accelerators. They are trapped within the strings and membranes.

Einstein's lesson

A smaller group of researchers than the string theory community are working in the other direction. They start from General Relativity and modify it in the hope of finding a quantum theory of spacetime. A number of researchers believe they have found how to do this. Their theory is called 'loop quantum gravity' and the practitioners in the field have been making steady progress over the last decade or two. Loop quantum gravity also predicts that space and time come ultimately in indivisible lumps down at the Planck scale. But there is a crucial difference between this approach and string theory. For in the latter, space and time are still regarded as an absolute background: the stage on which the strings perform their dance. That approach goes against one of the central tenets of Einstein's theories: that space and time can really only be defined in terms of the relationships between different events. Put simply, this would be something like the statement that the distance between two points only exists because the points themselves exist.

Loop quantum gravity starts from the correct picture of space and time. Its 'loops' are thus not physical entities like superstrings. Instead all that is real is the relationship between loops. Unfortunately, what loop quantum gravity cannot do is predict the properties of the four forces in a unified way.

Maybe more work has to be done to bridge the divide between quantum mechanics and relativity. Or maybe M-theory does indeed hold the key. We'll just have to wait and see.

Accentuating the Negative

Paul Davies Professor of Natural Philosophy, Macquarie University, Sydney

When Stephen Hawking announced in 1974 that black holes are not black, but glow with heat and slowly evaporate, people were left wondering where the heat energy came from. Since nothing is supposed to escape from a black hole, how could it supply energy to power the heat emission? The answer was soon found. The heat radiated by a black hole is paid for, not by energy flowing out of the hole, but by negative energy flowing in.

In simple terms, a negative energy state is one with less energy than a state that has zero gravitational field. But how can you have less mass, or energy, than completely empty space? The secret lies with quantum mechanics, or, more accurately, quantum field theory. According to this theory, apparently empty space is not really empty at all,

but teeming with all manner of 'virtual' particles that exist only fleetingly. The so-called quantum vacuum state cannot be divested of these countless ghostly entities, even though they have no measurable energy, and hence no gravitational pull, overall. Indeed, the virtual particles manifest themselves only when something disturbs the quantum vacuum.

One simple example of a disturbed quantum vacuum is the famous Casimir effect, discovered in 1948. Two mirrors placed face to face trap a slab of quantum vacuum between them. Mirrors reflect real photons of light, but they also reflect the ghostly virtual photons too. This upsets the energy budget of the quantum vacuum, and when the sums are done, you find that the total energy of the modified vacuum state between the mirrors is less than the unmodified one,

the one without the mirrors, i.e. empty space. So the presence of the parallel mirrors lowers the energy below what it is for empty space, making it negative by most people's definition. This negative energy manifests itself as a measurable force of attraction between the mirrors.

In the 1970s, Stephen Fulling and I discovered using quantum field theory that negative energy states could be produced by a single mirror, if it moved in a certain way. Furthermore our calculations showed that the negative energy would flow away from the mirror at the speed of light, opening up the prospect of a negative energy beam, as opposed to the static negative energy associated with the Casimir effect. Shortly after, it was found that mixed laser beams could create short bursts of negative energy, from what were known as squeezed states. Such states have recently been demonstrated in the laboratory.

Once the possibility of negative energy fluxes was established, Fulling and I were soon able to show that Hawking's black hole radiation was powered that way. At a distance, the heat radiation represents a flux of positive energy streaming away from the black hole. We knew, however, that this flux of energy could not be traced all the way back inside the hole, as that would violate the rule that nothing can get out. What we found is that a flux of negative energy continually streams into the hole from the region around it. As a result of the steady accumulation of negative energy inside the hole, its mass diminishes. Our calculations confirmed that the black hole loses mass-energy at exactly the rate needed to supply the outflow of heat radiation. Black holes, and in fact all spherical objects, create negative quantum vacuum energy in their vicinity because the curvature of spacetime due to their gravitational field disturbs the activity of the virtual particles. By one of those pleasing coincidences, the spacewarp created when a star implodes to a black hole turned out to have a mathematically identical effect on the quantum vacuum as an accelerating mirror.

The theoretical possibility of creating a flux of negative energy – in effect a beam of cold and dark, rather than heat and light – offered

some bizarre and puzzling scenarios. Suppose such a beam were directed at a hot object other than a black hole, such as an oven with an aperture protected by a shutter. It would seem that the contents of the oven would lose energy and cool down. But this would be a clear breach of the celebrated second law of thermodynamics, because the loss of heat by the oven would amount to a loss of entropy[1] too, and the second law forbids the entropy of a closed system from going down. (The beam itself has zero entropy.) The second law is the linchpin of thermodynamics, and any violation would open the way to a perpetual motion machine, which is not thought possible.

Yet another scenario involving negative energy that also leads to a paradox is that of a wormhole in space, made famous by Jodie Foster in the movie *Contact*. Wormholes are hypothetical tunnels or tubes of space that link distant points to create a shortcut. If they exist they can be used as time machines. There are circumstances in which an astronaut who transits the wormhole, exiting at some remote place and returning home via normal space, can get back home before he left! Thus the very existence of wormholes threatens us with paradox. Mathematical modelling by Kip Thorne and his colleagues at Caltech showed that wormholes are theoretically possible, but they can be transited only if some sort of negative energy state can be created in their throats. This is necessary because gravity threatens to crush the wormhole before anything can get through. Because negative energy has negative mass it exerts a negative gravitational force, opposing the pinching effect, and keeping the throat open. Thus, once again, unrestricted negative energy seems to lead to an unphysical paradox.

Although the foregoing paradoxes make theoretical physicists feel uneasy, nobody has yet proved that sustained negative energy fluxes or baths are impossible. It remains an open question whether, say, unlimited supplies of energy might lie untapped in the quantum vacuum.

[1] Entropy is a rather strange quantity and is the measure of the amount of disorder in a physical system. For instance, shuffling a pack of cards increases its entropy.

Chapter 9
Putting the
Quantum to
Work

I hope by now to have convinced you of how fundamental quantum mechanics is in underpinning so much of modern physics and chemistry. Of course, you might be hard-nosed enough to be thinking that this is all very fascinating stuff, but hardly relevant to the way you lead your day-to-day life. After all, the world of our experiences and senses is so very far removed from all that quantum weirdness going on deep down at the microscopic scale; surely it cannot impinge on us directly. So, in this chapter we investigate just how some of the quantum physics we have been studying has been harnessed to develop technologies over the past half-century that we take for granted today.

For instance, a certain property of electron wavefunctions – their obedience to the Pauli exclusion principle – explained how electrons arrange themselves in energy 'shells' in atoms, which led to an understanding of how metals conduct electricity. This, in turn, led to the study of semiconducting materials and the invention of the transistor, from which grew the microchip, the computer, and the Internet.

Furthermore, CD and DVD players make use of a corresponding

property of photons that gave us the laser, a device with all sorts of industrial, medical, and research, not to mention leisure and entertainment, applications.

The phenomenon of quantum tunnelling gave us nuclear power and will hopefully lead one day to a cleaner source of unlimited energy: nuclear fusion.

Yet another strange quantum property of certain materials that are cooled to near absolute zero gave us superconductors, which might one day provide us with the ultimate solution to energy conservation: power cables with no electrical resistance.

And you thought radioactivity was bad for you? What if I told you it has revolutionized medicine? And how do you think your smoke detector works?

The list is very long, but I will focus on just a few of the more important technologies that directly rely on quantum mechanics.

The age of the microchip

Nowadays it seems as though everything from cars to washing machines, coffee makers to musical birthday cards, contains a microchip inside. But have you ever wondered just how a chip works?

In the box on the Pauli exclusion principle in Chapter 7, I described how electrons arrange themselves inside atoms such that no two electrons are allowed to occupy the same quantum state. That is, they cannot be described by the same wavefunction but must differ in some way: either in their energy, their orbital momentum or their spin direction. This is because electrons, along with the other basic building block of matter – the quarks – belong to the class of elementary particles called fermions (named after the great Italian physicist Enrico Fermi). Fermions are said to obey the exclusion principle and prefer to keep themselves to themselves, each in its unique quantum state. Force-carrying particles, such as photons, belong to the other class called bosons, which are far more sociable and have no problem with, indeed prefer, having identical

Increasing energy

Electron energies in solids. Each shaded well signifies a single atom in a lattice. Many electrons are tightly bound within their host atoms, occupying discrete quantum orbitals, and therefore do not move around the lattice. However, the higher energy levels occupied by the outer electrons merge together to form 'energy bands', each with a continuous range of allowed energies. In between these bands are energy gaps that the electrons are forbidden to occupy. In a conducting metal (top) the highest (valence) band is only partially filled with electrons which are therefore free to move around the lattice, and hence to conduct electricity. In an insulator (middle) the valence band is completely full and the quantum rules forbid the electrons from moving around. Since the gap to the next band is too great to jump across they are stuck. In a semiconducting material (bottom) the valence band is also full, however the gap to the next band is very small and a few electrons are able to jump up, creating a new conducting band.

wavefunctions and occupying the same quantum state. The exclusion principle does not apply to bosons.

The rules of quantum mechanics have thus provided us with an understanding of atomic structure. The outermost electrons in an atom define its chemical properties, which in turn explain how atoms bind together to make different materials. When atoms stack together in bulk matter, the outermost electrons find that they have a little more freedom. Instead of occupying a certain specific energy level, they span a broad range of energies called the valence band. Indeed, due to the large number of atoms involved, the allowed energies in this band are so closely spaced that it can be regarded as having a virtually continuous spread of energy.

In certain materials, such as metals, this valence band is only partially filled and the Pauli principle does not restrict the electrons' movement. When a voltage is applied across a metal, these valence electrons move through it freely, forming an electric current. The valence band is referred to as the conduction band in such materials, which are called electric conductors.

If, however, a material is made of atoms in which the valence band is full, then all states in the band are occupied and the electrons have nowhere to go. The electrons in the valence band can no longer slosh about freely and we do not refer to it as a conduction band.

There is generally a gap before the next energy band begins. If this gap is too large then the electrons just stay where they are since they are unable to leap up to freedom. And if they cannot reach this band they will be unable to flow through the material when an electrical potential difference is applied. Such materials are insulators.

Materials with atomic structures in between conductors and insulators have a rather interesting property. The atoms in a lattice of silicon, for instance, have a valence band that is full of electrons, but the energy gap between this and the band just above it is bridgeable, so a few of the more energetic electrons can leap up to the higher band, turning it into a conduction band. Such materials are thus 'semi-conducting'. An additional subtlety is that if an electron in a semiconductor does escape from the valence band to freedom, it

will leave a hole behind it that an electron in a neighbouring atom can occupy. This in turn leaves a hole in the neighbouring atom's outer band. Thus an electric current that pushes the electrons in one direction will also cause the positive hole[1] to move round in the opposite direction, so you get double the electrical effect.

In pure silicon, the above process happens only rarely since an excited electron will quickly fall back down again, either into its original position or into a hole in a nearby atom. At room temperature only one in every ten trillion silicon atoms will contribute a conducting electron. This number can be increased by raising the temperature as this provides extra energy to excite more electrons.

But another way of significantly increasing the conductivity of a semiconductor is by adding tiny amounts of a different element, called impurity atoms. There are two types of impurities, depending on whether their atoms supply extra electrons to enhance conductivity or remove silicon electrons, to create holes, which also enhances conductivity. The two types of semiconductors that are created in this way are known as 'n-type', since they add negative charges, and 'p-type', since they increase the number of positive holes by stealing electrons.

The reason all this is important is that when n-type and p-type semiconductor strips are put together, they form an electrical 'switch', known as a p-n junction, that controls current flow in a very clever way. Such a device is called a 'diode' and has the property that it allows current to flow across it much more easily in one direction than in the opposite direction. The basic principle is as follows: when the two types of semiconductors are brought into contact, some of the free electrons in the n-type migrate across to cancel the holes on the p-type side. This causes a thin layer to build up at the boundary, which is negatively charged on the p-type side (since it has lost its positive holes) and positively charged

[1] It is only positive in the sense that it is missing an electron, so it is positive relative to the surrounding electrons.

on the n-type side (because it gave up some of its electrons). This 'depletion' layer acts like a one-way street for current flowing across it by acting like an electrical valve: it allows current to flow far more easily one way (from p-type to n-type) than the other way around.

Such semiconductor diodes are used for a whole host of different applications. For instance, they can be made to emit light when current is passed across the junction, forming a 'light emitting diode', or LED, as it is commonly known. But by far the most famous use of semiconductors is to form a 'transistor'.

The very first transistors were invented soon after World War II by three Americans, John Bardeen, Walter Brattain, and William Shockley, while working for Bell Labs. The basic idea is to use two p-n junctions back to back. Depending on which way round this is done, you can get two types of what is called a bipolar junction transistor having either an n-p-n (a p-type sandwiched between two n-types) or a p-n-p (n-type between two p-types) arrangement. A high current flowing the high-resistance (difficult) way through one of the junctions can be controlled by a much smaller current flowing the easy way through the other. The word 'transistor' comes from combining the two words describing the processes in this device: transfer-resistor.

Transistors, with their ability to amplify currents and voltages across their terminals, turned out to be the perfect devices to make up logic gates, the fundamental units of a computer. They operate according to the rules of binary logic of zeros and ones, being either 'on', if current flows through them, or 'off' if no current flows.

Transistors form a part of electric circuits containing various other electrical components such as resistors and capacitors. In the late 1950s, it was realized that such a circuit could be etched out of a single block of semiconductor material, forming what is called an integrated circuit. Small pieces, or chips, of silicon wafer were used for this purpose, hence the popular name.

Soon after, such chips began to be mass-produced and quickly replaced the cumbersome vacuum tube valves that had been used in

the first generation of computers of the 1940s and 50s. By the late 60s, each chip could incorporate thousands of transistors in its integrated circuit. Today's most powerful microchips have literally millions of transistors squeezed onto an area the size of your fingernail. And what's more, such 'microprocessors' are not simply hard-wired to perform a specific task, but can be programmed to carry out a multitude of functions.

While we have quantum mechanics to thank for the technology behind the ubiquitous microchip, the way electrons arrange themselves in atoms, which gives rise to semiconducting materials, may not sound to you like particularly weird quantum behaviour. I should therefore mention another device also used in modern electronic circuits, which exploits quantum mechanics more directly. The tunnel diode uses the fact that electrons can jump across an insulating layer that they have no right being able to traverse at all. But don't be tempted to think of them as tiny particles hopping across a wall in some classical sense; more correctly, it is the electron wavefunctions that leak across the insulating region. And with this quantum leap they can cover the distance far more quickly than they would normally move through a transistor. Therefore such devices are used as very fast switches in microprocessors.

A clever idea looking for a use

That was how physicists regarded the laser when it was first invented in 1958. Today, the laser has a wide range of applications, from shipbuilding to eye surgery to CD players to supermarket checkouts. But the principles behind it involve more than just the production of an intense beam of light. The physics of the laser is pure quantum mechanics. Unlike the microchip, which relies at its heart on the properties of electrons and their strict adherence to the law of the exclusion principle, the laser is dependent on the camaraderie and co-operation exhibited by photons.

When an electron gains energy by absorbing a photon it can

jump up to a higher atomic orbit. The difference in energy between the two orbits will be equal to the energy of the absorbed photon, which in turn depends on its frequency according to Planck's formula. The 'excited' electron will shortly after spontaneously drop back down again by emitting a photon with the same energy. This process is called spontaneous emission and is the way an electric light bulb works. The current flowing through a tungsten wire heats it up, causing the electrons in the tungsten atoms to gain energy and become excited to higher orbits. When they drop down again, they emit photons over a wide spectrum of frequencies including those in the visible light range.

A laser works differently. If, instead of the electron dropping down to its original state of its own accord, it is stimulated into doing so by being knocked by an incoming photon, then we get two photons coming off: the original one and the new one emitted by the electron. The two photons can then each prompt further excited electrons to drop down, releasing yet more photons, a bit like a chain reaction. This process is called stimulated emission and is where the laser gets its name: Light Amplification by the Stimulated Emission of Radiation.

The released photons, being bosons, come off in the same quantum state as the incident photons; they have the same wavelength, are in phase, and travel in the same direction. In this way, laser light is said to be coherent.[2] It can have very high intensity and can be focussed into a narrow beam.

The very first laser was built in 1960, and a wide range of uses has since been found for it. Lasers can be designed to weld, cut, and melt. You find them on car production lines and in the clothing industry. They can be used to check alignment of very straight trenches or to fit together large components in heavy industry with high precision. Their precise wavelengths can be tuned and utilized in fields such as interferometry to measure lengths to exceptional accuracy. (For

[2] As opposed to the light from a standard light bulb, which emits incoherent light since the waves streaming out are not in phase or even of the same frequency.

Unlike normal light from a torch, which is a mix of electromagnetic waves of many different wavelengths spreading out in all directions, laser light is highly organized. All the waves have the same wavelength, corresponding to the characteristic frequency of the emitting atoms. Laser light is also of high intensity due to the amplification process and does not spread out after it is emitted, which allows it to be beamed over very long distances.

instance, lasers have been used to measure the distance between the Earth and the Moon to an accuracy of a few centimetres.) They are also used to create holographs: three-dimensional images with a variety of applications including, possibly, incredibly effective information storage devices.

Lasers can also be made from semiconductor diodes in a similar way to LEDs. These cheap and cheerful solid state lasers are robust, reliable, and the size of a grain of sand. Nowadays, such devices are used in communications to send light down optical fibres, and are found in CD and DVD players, and supermarket checkouts where they scan bar codes.

Due to their ability to focus energetic photons with such remarkable precision, lasers have also become crucial in the computer industry. They are used in a process called photolithography to etch the cuts and grooves of integrated circuits onto silicon chips.

Laser fibre-optics will soon become the normal way information is transmitted around the world, and is where the future of the Internet lies. We are almost at the stage where our interactive TV and PC will be one and the same. Indeed, it is already possible to download movies from the Internet and watch them on your PC. Fibre cables transmitting laser light will soon be able to carry billions of bits of information, equivalent to the full works of William Shakespeare, in under a second![3]

What is possibly most remarkable of all is that lasers can nowadays be used to control single atoms in ways that are opening up a whole new field of quantum technology that will be discussed in the last chapter.

Magnets as big as houses

While I consider myself a reasonably competent handyman around the home, I do tend to get, along with most other theoretical physicists around the world, the usual good-natured ribbing from my experimental colleagues about my inadequacies in the laboratory. And while it is true that I manage to completely avoid any sort of experimental work, I had not realized until recently just how oblivious I had become to the technology that went into producing the experimental results I try to understand. However, that is not the point of the story I am about to tell, but rather to highlight another quantum application you may have heard of.

During 1999, I spent six weeks at Michigan State University's cyclotron facility where I worked on interpreting new data from

[3] For a breathtaking tour of the future of microelectronics and telecommunication, not to mention other areas of science that will revolutionize the 21st century, I recommend Michio Kaku's book *Visions* (Oxford University Press, 1998).

their most recent experiments on 'exotic' nuclei.[4] The National Superconducting Cyclotron Laboratory at MSU is one of the world's leading nuclear physics research laboratories, in which rare and exotic species of nuclei are created and studied. The 'cyclotron' bit means that these nuclei are accelerated up to high energy inside a powerful magnetic field before they are smashed deliberately against a nuclear target to see what happens. Cyclotrons utilize powerful electromagnets that force charged particles or ions to travel around inside them, spiralling outwards and gaining in speed as they go.

I knew this much before I went to MSU, and of course I also knew the basics of what happens to the charged ions once they leave the cyclotron and enter the beam analysis system on their way to the target area. In fact, for my research purposes, I could get away with remarkably sketchy information about all the finer details. It was not the raw data, or knowledge of the ingenious techniques (and believe me there are many) that went into getting it, that I needed to know, only the final fully processed data.[5]

The accelerator facility is housed in a large building in the middle of campus and has office space around two sides. During my stay I would enter the building through one of several doors and make my way to the office where I would spend my time developing the mathematical models of the nuclear reactions being studied, writing computer code, or just discussing physics with the other theorists there. I would of course regularly go down the corridor for coffee, for discussions with my experimental colleagues, or to attend seminars.

One day, towards the end of my stay there, I was discussing something with a colleague, Gregers Hansen, in his office. Gregers is a leading nuclear physicist, originally from Denmark but settled in the US. He was one of the discoverers of the nuclear halo that I

[4] In this context, an exotic nucleus is one that has a dramatic imbalance between the number of protons and neutrons it contains. This not only makes it highly unstable but often endows it with strange new properties.

[5] I am not proud of this attitude and have since tried a little harder to appreciate the experimental details.

described in a box in Chapter 3. I noticed his computer screen kept flickering and he told me he had to keep 'degaussing' it to neutralize stray magnetic fields. Conveniently enough, there was just such a button on his computer monitor. I asked why this was necessary and he explained that it was due to the testing that was being carried out on the cyclotron. When he saw the puzzled look on my face and realized my ignorance he elaborated that the problem was due to the magnetic field of the larger of the two superconducting magnets that had been switched on that day. When I pointed out that I had not actually seen these he was amazed. 'You mean to say you've spent six weeks sitting less than ten metres away from the world's most powerful cyclotron magnet[5] without realizing?' The next day, the last of my stay, he took me on a grand tour of the laboratory. Sure enough, on the other side of the wall from my office, and through several metres of shielding from radioactivity, was a magnet so big that we had to climb stairs around its outside to get to the top. I have since spent some time at TRIUMF Lab in Vancouver which contain's the world's largest cyclotron magnet, though less powerful than the MSU one.

What is of interest here is the way such magnets work. For they make use of another purely quantum effect: superconductivity.

Perpetual motion electricity

The reason conductors all exhibit electrical resistance is that the conducting electrons continually bump into the vibrating atoms

[5] The larger of the two cyclotrons, the K1200, accelerates atomic nuclei to one-third of the speed of light (100,000 kilometres or 61,000 miles per second). Each nucleus will cover a distance of about three kilometres having travelled around in a circle about 800 times. The magnetic field felt by the nuclei as they exit can be up to five Tesla. To put this in perspective, the magnetic field strength of the Earth is one ten-thousandth of a Tesla, and a typical bar magnet is around a tenth of a Tesla at its poles. Taking into account just how tiny and light a single atomic nucleus is, we see that a magnetic field of several Tesla can have a pretty dramatic effect on it. For the record, the highest magnetic fields, of up to 60 Telsa, are generated using special 'pulsed' magnets.

of the metal. These vibrations increase as the metal is heated, thus increasing the resistance. Sure enough, by cooling a metal down we reduce the atomic vibrations and hence the electrical resistance. But in 1911 a very surprising effect was discovered.

When mercury is cooled to below 4.2 Kelvin[6] its electrical resistance suddenly drops to zero and it becomes a superconductor. We now know of many metals and metal alloys that exhibit this phenomenon when cooled to below their critical temperatures, typically a few degrees above absolute zero. This means that a current flowing through a superconductor never diminishes and requires no voltage source to sustain it once it is flowing. And while we cannot be sure that the resistance in a superconductor is exactly zero, it is as near to zero as makes no difference.

The origin of this effect is quantum mechanical. The Nobel Prize was awarded to the three people who, in 1957, explained how it works: John Bardeen – already a recipient of a Nobel Prize for his part in the invention of the transistor – Leon Cooper, and Robert Schrieffer. At a certain temperature, electrons with similar wavefunctions tend to pair up into what are known as 'Cooper pairs', named after one of their discoverers, due to a very weak attractive force. This has its origin in the way the surrounding environment is affected by the presence of each electron. (I will skip any further technical detail here.)

Each Cooper pair then behaves as a single boson (two fermions coupled together always make a boson).[7] Of course, you would be forgiven for thinking that the electrons in these Cooper pairs move

[6] The absolute zero of temperature is at 273°C below freezing point of water (–459°F), This is referred to as zero Kelvin (0 K). So we can say that water freezes at 273 K. We can never quite reach absolute zero but can get to within a small fraction of a degree of it.

[7] This is to do with the spin of fermions, which is always an odd multiple of half integer values, such as 1/2, 3/2, etc. Two spin-1/2 electrons will, if spinning the same way, combine to make a total spin of 1, and, if spinning in opposite directions, 0. Either way, a quantum system with an integer (0 or 1) spin combination will exhibit bosonic behaviour.

around stuck together like Siamese twins. After all, if they make a boson then surely this is like a new 'particle'. But we must not forget our quantum lessons. The two electrons need not be in close proximity. In fact, the distance between each member of such a pair is a thousand times greater than the average distance between any two individual electrons in the material! We say that the Cooper pair is in an entangled state described by a single wavefunction. Their behaviour as a single boson is thanks to their nonlocal connection. The two electrons in a Cooper pair are in continuous communication, like teenagers in different parts of a school playground chatting to each other on their mobile phones – which is not a quantum effect but just as strange!

Finally, there is no electrical resistance since, for this to happen, one of the two electrons would have to knock into, and scatter from, an atom hard enough to break up the pair. Therefore, any such scattering has to be strong enough to do this. At very low temperatures, the binding force between Cooper pairs is strong enough to withstand the weak atomic scattering. Hence the electrons in the pair simply do not collide with the atoms and there is no electrical resistance.

One of the main applications of superconductivity is, as in the case of the Michigan State University cyclotron, to make powerful magnets. (My story was an example of how quantum weirdness has been utilized to build a machine to study further quantum weirdness, found in exotic atomic nuclei!) A normal electromagnet involves passing a current through a wire coil, which generates a magnetic field. The more loops in the coil, the stronger the field. However, this also means higher resistance that weakens the current, and so a larger voltage is needed to compensate. This problem disappears if the coil is made from superconducting material. In practice, large conventional electromagnets have an iron core inside the electric coil, since without it the amount of current needed would make the magnet hopelessly expensive. But with an iron core the magnetic field strength saturates at some maximum and the only way to get more 'bending power' is to have a larger magnet (more iron). For

such a magnet to do the work of the MSU superconducting magnet it would have to be as big as a medium sized apartment complex!

Today, superconducting magnets have a wide range of applications. For example, they are used as magnetic separators in the mining industry and as magnetic navigation systems in medicine where they allow surgeons to guide catheters through the body, deliver drugs, or carry out biopsies.

Just as photons in a laser act in harmony to amplify a quantum effect to macroscopic levels, so too are the Cooper pairs in a superconductor able to co-operate as a group of bosons. It's a little like the large schools of fish swimming as one entity that we are used to seeing on TV wildlife programmes. One application of this behaviour is in a device called a SQUID (which is an acronym for 'Superconducting QUantum Interference Device'). SQUIDS contain components called Josephson junctions that are somewhat like tunnel diodes. But instead of two semiconductor strips, a Josephson junction is made up of two superconductors separated by a thin insulating layer. The Cooper pairs can tunnel through this layer easily making the SQUID incredibly sensitive to minute changes in magnetic fields. One application of such a device is in medicine, where brain activity can be studied by monitoring the magnetic fields produced by tiny electrical currents associated with individual neurons.

Finally, one of the most active areas of physics research over the past fifteen years has been in the field of high temperature superconductivity. It is all very well having a superconducting material, but it must be maintained at such low temperatures, often using liquid helium,[8] that it becomes inconvenient for more practical uses. In 1986, certain ceramics were discovered that are superconducting all the way up to temperatures of over 100 K. (This is still 173°C below freezing!) Current research in this field seeks to address three problems: whether there are materials

[8] Helium is in its liquid form below 4 K (–269°C or –452°F).

that are superconducting at room temperature, whether any of these materials are malleable metals, since ceramics are brittle and powdery and so not suited to being stretched into electrically conducting wires, and, finally, what exactly causes such materials to behave in this way at such (relatively) high temperatures.

If we do ever succeed in making electric power cables from room temperature superconductors it would help towards providing cheaper electricity. Current energy efficiency from electric power is low due in part to the energy lost as heat from the electrical resistance in the transmission cables that criss-cross the land. With superconducting cables we could transmit up to five times the power much more efficiently, and could thus afford to cut down dramatically on the use of polluting fossil fuels.

Energy from nuclei

Since I have strayed onto the subject of energy needs, now is a good time to describe another application of quantum mechanics. Nuclear power currently produces one-sixth of the world's electricity (rising to almost two-thirds in France). The basic idea behind nuclear power is the fission reaction, whereby heavy nuclei absorb neutrons, causing them to split into two pieces and release energy. In the process they produce further neutrons that get absorbed by neighbouring nuclei, which also split, thus maintaining a chain reaction. The heat produced in these reactions is used to turn water into steam that drives turbines, generating electricity.

A cleaner method of extracting the energy locked inside the atomic nucleus is through the reverse of this process. In a fusion reaction, two light nuclei are pushed together to make a larger, more stable nucleus. This is accompanied by the release of a huge amount of energy that can be harnessed. Such 'thermonuclear fusion' energy is the source of the heat and light from the Sun, and is the reason all stars shine.

Starting with two hydrogen nuclei (single protons) a sequence of

steps ends with the formation of a helium nucleus (two protons and two neutrons: an alpha particle). The problem is that two protons will repel each other unless they can be pushed very close together – in effect, quantum tunnelling through the energy barrier between them. For this to happen we require very extreme conditions of high temperature and density; no problem inside the Sun, but rather difficult to maintain in a laboratory on Earth.

An interesting point to make here is that the first step in the fusion process within the Sun involves the combined participation of all four forces of nature. The gravitational force of attraction between all matter ensures that the hydrogen is packed close enough together for two protons to stand a good chance of fusing, whereby the attractive strong force wins over the repulsive electromagnetic force to push them even closer together. But since two protons do not bind permanently together, whereas a proton and neutron will, one of the protons is forced to undergo beta decay and change to a neutron, which is where the weak nuclear force comes into play. Once this has happened, we end up with a bound proton and neutron.

A nucleus made up of just a proton and neutron is known as a deuteron, since it is a nucleus of the heavy isotope of hydrogen, deuterium. The deuteron has played an important role in the development of nuclear physics. It is the simplest composite nucleus – any smaller and we are left with a single proton or neutron – and so provides a testing ground for the properties of the nuclear force that holds nucleons together in nuclei. It is also a very useful probe that can be made to react with other nuclei in order to study their properties. I was fortunate enough to have as my PhD supervisor Ron Johnson, one of the world's leading experts on the structure and reaction properties of the deuteron, and known to his students inevitably as Deuteron Johnson.

Once deuterons are formed in the Sun they can fuse with further protons to form helium nuclei. This requires the deuterons and protons to tunnel through their mutual Coulomb barrier.

Fusion research, involving deuterium and tritium (an even

heavier isotope of hydrogen whose nuclei are composed of a single proton and two neutrons) is being carried out by many countries. It is currently at the stage where the 'break even' point has now been surpassed, whereby the amount of energy being produced as a result of the fusing process is more than the energy that is put in to create the necessary starting conditions that push the fusing nuclei together. The next step is to sustain the fusion conditions for more than just a fraction of a second!

Not only is there an unlimited supply of fuel for fusion reactors (water) but energy from fusion produces far less harmful waste products than nuclear fission. This is why most of the wealthier nations have invested, and continue to invest, so much in this research despite the slow progress.

Quantum mechanics in medicine

There are far too many applications of quantum mechanics in medicine for me to list here, ranging from X-rays to laser surgery. I will therefore briefly discuss just two. One utilizes the quantum mechanical spin of nuclei while the other involves the use of antimatter to map the brain.

The principles of the technique of Nuclear Magnetic Resonance (NMR) were developed by American physicists half a century ago and were, for many years, used as a spectroscopic tool in chemistry. The idea is to measure the concentrations of different atoms in materials from the radiation given off by their nuclei as their quantum spin is flipped around in a magnetic field. Many nuclei rotate (quantum mechanically) due to the combined spin of their constituent protons and neutrons. The axis about which a nucleus spins will, like a compass needle, align itself with the lines of an externally applied magnetic field. Depending on the direction of the spin (in a loose classical sense) we can define this axis as either pointing one way or the other along the magnetic field lines. Such spin directions are referred to simply as 'spin up' or 'spin down'. Some of the nuclei

are then induced to flip their spin over by the application of a high-frequency signal from an oscillating magnetic field. When they return to their original orientation they release a tiny burst of energy specific to that type of nucleus.

Most materials, such as the organic matter making up our bodies, are composed of molecules, which in turn comprise different sorts of atoms. Using a technique known as 'projection construction', NMR signals from the continuous flipping-over of the spin of certain atoms can be used to generate images of the interior of the human body. The method is quite harmless since, unlike in X-ray tomography, the body is not penetrated by harmful ionizing radiation. Despite this, the machines used are commonly known as MRI (Magnetic Resonance Imaging) scanners, due to public fear of the context of the word 'nuclear' in NMR.

Positron Emission Tomography (or PET, as it is more commonly known) is a wonderful technique that is relatively easy to explain. One particular use for a PET scan, for instance, is to map the brain of a stroke victim or that of a newborn baby if there has been a danger of oxygen starvation during the birth. First, glucose containing harmless radioactive isotopes of carbon or nitrogen is injected into the body, whereby it is carried to the brain. Regions of high and low neural activity can then be mapped by measuring the concentration of glucose, the brain's energy source, in different regions.

The radioactive isotopes in the glucose undergo beta decay and emit positrons. These particles almost immediately encounter electrons and the process of pair annihilation takes place in which the electron and positron disappear in a flash of light with two energetic photons being emitted back to back. These photons are picked up by detectors known as photomultipliers and their paths traced back to the precise position in the brain of the electron/positron annihilation. The decayed radioactive nucleus and hence the glucose that was carrying it is thus located. By detecting a large number of such photon pairs over a period of time, a series of images can be created that map out the changes in glucose concentration.

atom

gamma ray
photon

Fluorescent lig
(visible photon

photomultipliers

photosensitive screen

electron

measuring device

e⁻

e⁺

Oxygen-15

How a PET scanner works: Atoms of an unstable isotope are used to track glucose molecules in the brain. The nuclei of these atoms beta decay by emitting a positron (the antimatter partner of the electron). This quickly collides with, and annihilates, an atomic electron, releasing two back-to-back gamma ray photons. A PET scanner consists of arrays of photon detectors which capture the gamma rays in scintillator crystals. Many atoms in these crystals are excited by a single high energy gamma ray but quickly settle back down by releasing lower energy photons. These in turn trigger the release of a cascade of electrons inside a photomultiplier tube, ending in an electrical pulse output. These signals are used to build up a map of the brain, slice by slice, depending on where the higher concentrations of glocuse are. While being able to look inside the human brain using radioactive atoms that spit out antimatter might sound pretty scary, PET scanners are in fact less harmful than X-rays.

Quantum mechanics and genetic mutations

A few years ago, a microbiologist colleague of mine, Johnjoe MacFadden, and I published a speculative paper in the American journal *Biosystems* in which we proposed a quantum origin to an unusual type of genetic mutation, known as an adaptive mutation, of *E. coli* bacteria. It was speculative for two reasons. Firstly, the process of adaptive mutations was, and still is, subject to some controversy. Secondly, the quantum mechanism we were proposing required a certain unique quantum property of living cells, otherwise our theory would not hold water.

It is now known that the genetic coding properties of DNA molecules in all living cells are due to the nature of hydrogen bonds between base pairs. Ever since Francis Crick and James Watson discovered the double-helix structure of DNA, it has been known that certain natural mutations can occur due to the random quantum tunnelling of protons from one site on the DNA to a nearby one, forming a different chemical bond. This sort of accidental error in the coding of the DNA will occur in one out of a billion sites, but when it does we get a quantum mutation. So quantum mechanics certainly has some part to play in evolution.

In adaptive mutations, no such straightforward explanation can be found. When a strain of *E. coli* cells, called *lac-*, that is deficient in an enzyme that allows them to feed on lactose, are provided only with lactose, most are expected to starve. A few, however, should randomly mutate into the strain *lac+*, which can feed on the lactose and grow, and hence begin to replicate. What is found instead is that many more *lac- E. coli* mutate to *lac+* in the presence of lactose than when none is present, even though they

can have no knowledge that the lactose is present until after they mutate. It seems like magic.

Here is where the idea of quantum superposition might provide an explanation. You see, in each cell the mutation from *lac-* to *lac+* might be due to the tunnelling of a single proton between two adjacent sites. Quantum-mechanically of course the proton's wavefunction is such that it has a certain probability of being found in either site: a superposition of tunnelled and not tunnelled. And, provided such quantum coherence can be maintained within the cell for long enough, the whole DNA should evolve as a superposition of mutated and unmutated states.

The problem of course is decoherence, the leaking away of quantum weirdness into the surrounding environment. This should take place over a time scale of less than a billionth of a second, far too quickly for the necessary changes within the cell to take place in order for it (or a part of its wavefunction) to mutate to the strain that is aware of and can utilize the lactose in its environment. However, we wanted the superposition to persist long enough for the lactose itself to cause the decoherence. That is, the lactose does the job of opening the box containing Schrödinger's cat and collapses the state of the cell into one or other state. This is because, given the right *E. coli* strain, chemical reactions take place involving the lactose, which can also bring on decoherence. If the decoherence in the presence of the lactose takes place more quickly than in its absence then we could explain the surprising results seen in the adaptive mutation experiments.

The way this could happen would be due to a simple mechanism whereby repeated measurements 'drag' a quantum system from one state to another. If the lactose 'measures' the cell to be in the *lac+* mutated state, which would be rare, then the cell will go ahead and feed and grow. But if the cell is in the *lac-* state then nothing happens and it reverts back to being in a superposition of the two states again. Therefore, the more frequently the measurement can take place the more regularly the *lac+* mutations can be harvested.

Unfortunately, it is hard to believe that quantum coherence would persist for so long in such a busy and complex environment as a living cell. The only possibility would be if a cell behaves in a way that is very different to an inanimate system of equivalent size, complexity, and temperature. Indeed, it was suggested by Erwin Schrödinger in his famous 1944 book *What Is Life?* that living cells maintain a structure and order that is equivalent to normal matter at close to absolute zero, a regime where quantum effects persist for much longer. But since no one really understands what it is that makes life so special, our theory remains speculation.

Since publication of our paper, I have gone somewhat cool on the idea. However, physicist Paul Davies, one of the few who made encouraging noises about our proposal, is also interested in such questions. He points out that we have yet to understand the nature of decoherence properly, not to mention the complex quantum properties of living cells, and much work has still to be done.

Microscopes to See Atoms

By the early 1930s, the limit of light microscopes – those that rely on focusing the light from a sample through lenses – had been reached. The wavelength of visible light limited the maximum magnification to about 1000x (and the resolvable detail to fractions of a micron). The desire to see finer detail than this in living cells drove scientists to consider using waves of much shorter wavelengths. Here is where the quantum wave nature of electrons came in useful. The matter waves associated with beams of electrons have wavelengths far shorter than those of visible light. The first electron microscope was developed in Germany in 1931, and used a focused beam of electrons to study the structure and composition of a specimen. Such a Transmission Electron Microscopes (TEM) works on the same principle as a slide projector, whereby the part of the beam transmitted through the specimen is projected onto a screen.

After the Second World War, a new type of electron microscope, known as a Scanning Electron Microscope (SEM) was developed. It involved scanning an electron beam across the surface of the specimen. An image of the surface is produced by collecting and amplifying the electrons that bounce off or are knocked out of the specimen.

Nowadays, microscopy has developed to the point that resolution and detail is down at the level of individual atoms. The Scanning Tunnelling Microscope (STM), first developed in the early 1980s, involves dragging a sharp conducting metal needle above the surface of the specimen. Electrons are able to quantum tunnel across the gap onto the surface, and the electric current thus produced can be measured

and related to the size of the gap, and hence the positions of the atoms on the specimen's surface.

The principal drawback of the STM is that the specimen must also be a very good electrical conductor, and samples have to be specially prepared. By 1989, the first commercially available Atomic Force Microscope (AFM) was introduced. While still employing a similar scanning technique involving a sharp tip to probe and map a surface, such microscopes did away with tunnelling electrons and instead made use of a very sensitive spring that pushed the tip very gently onto the surface. The force used to keep the tip in place as it bumps its way around the surface is of the order of inter-atomic forces (about 10^{-9} of a Newton), hence its name.

Today, scanning probe microscopy, using both STMs and AFMs, is used for more than just imaging surfaces. Physicists and chemists are able to use the probe to move molecules, and even individual atoms, around in a controlled way. New techniques are currently being developed and applied to fields ranging from biology to semiconductors, data storage to the study of atomic structure in polymers and crystals. In the future, STM tips may even be used to build a whole array of tiny molecular machines, the Holy Grail of the burgeoning field of nanotechnology.

The tip of an STM scans the surface of a material by allowing electrons to tunnel across the gap, producing an electric current that provides a measure of the contours of the surface. It can even be used to push atoms around the surface in a controlled way.

Atomic engineering and nanotechnology

As its name suggests, nanotechnology is the relatively new field of building and utilizing micro-machines on the scale of nanometres (10^{-9} metres or millionths of a millimetre). This is the scale of atoms and molecules and is comfortably within the quantum domain. Many physicists now believe that there is no reason why such tiny molecular robots, comprised of circuits, cogs, levers, and gears, cannot one day be built. To give you an idea of the scale of these entities, they would be to an ant what an ant is to a football stadium. Already, scientists have shown that they are able to manipulate individual atoms and molecules in a controlled way and many feel that creating such nanoscale designer robots is not too far away. It is predicted that nanotechnology is set to spark an industrial revolution that will transform our lives to a greater extent than even silicon microelectronics did in the last century.

One possible futuristic development might be that such molecular robots would be able to replicate themselves, not through a biological process but by simply building copies of themselves, in trillions, out of the raw material provided in their surroundings. In this way, whole armies of such machines could be utilized to carry out a bewildering array of tasks, from killing cancer cells (one by one!) to performing miraculous feats of engineering and even creating what are being dubbed 'smart materials'.

While such technology is still decades away, the first few steps have already been taken. In the early 1990s, researchers in Japan were studying soot (produced in an electric discharge between two carbon electrodes) in a transmission electron microscope.

They discovered tiny molecular carbon tubes that have since been found to have amazing properties. Such carbon nanotubes are nothing more than rolled up single sheets of graphite (known as graphene), one atom thick. Due to the properties of the carbon bonds, nanotubes could be used as the ultimate high-strength fibre, with applications from tips for scanning probe microscopes to earthquake-resistant buildings. Cars made of carbon nanotubes would unbuckle to their original shape after a crash.

Nanotubes can also act as molecular wires that can be designed to be either fantastic electrical conductors (almost as efficient as superconductors) or as semiconductors to be used in integrated circuits. They could even be used to efficiently transport individual atoms and molecules around.

Watch this space.

Bose–Einstein Condensates

Ed Hinds Professor of Physics, University of Sussex

Imagine a single atom trapped in an exceedingly small bowl. A quantum mechanical wavefunction describes the movement of the atom within the trap. The wavefunction vibrates by different amounts in different places and the square of the amplitude of vibration determines the chance of finding the atom there. If we suppose that the atom can't penetrate into the wall of the bowl then the wavefunction has to be zero there.

The figure above shows the three simplest possibilities – wavefunctions with 1, 2, and 3 zones of maximum vibration. The shorter the distance between zeros of the wavefunction, the higher the energy of the atom, so these three wave possibilities correspond to the lowest three possible energies of an atom in the bowl. The various energy states are labelled by an index $n = 1, 2, 3$, etc. known as a quantum number. If a cloud of many atoms is held in the bowl, any particular one is jostled around and spends its time in a superposition of many different quantum states. The higher the temperature, the bigger the range of quantum states being sampled. The ground state is rarely occupied under normal conditions because there are so many other possible states.

But if the gas becomes cold enough and the density of atoms is high enough, it can happen that the ground state starts to be occupied by two atoms at once. If they are

the right kind of atoms (bosons) the result is quite spectacular. The wavefunctions describing the two atoms add up to make one with double the vibration amplitude and this quadruples the probability of finding an atom there – in other words more atoms are attracted into the state. The more atoms go into the ground state the more the wavefunction amplitude grows and the more attractive that state becomes to other atoms being jostled about in the gas. This is an interference effect due to the constructive addition of the waves describing each atom. The result is an avalanche in which nearly all the atoms agree to occupy the lowest energy state, while a small number of them take up all the rest of the energy in the gas. These are often so energetic that they escape from the bowl entirely, leaving nearly every atom in the ground state. This process is called Bose-Einstein condensation and the ensemble of atoms all in the same quantum state is a Bose-Einstein condensate or BEC [1].

A BEC has remarkable properties due to the fact that all the particles are behaving identically. In liquid helium, about 10 per cent of the atoms form a BEC but that part of the liquid flows without any viscosity, which makes it a superfluid. The light inside a laser is a BEC of photons (photons are bosons), which means that the photons in the laser beam have identical behaviour – hence the pure colour and directionality of the beam. Bose-Einstein condensation of atomic vapour was first seen in 1995 in clouds of rubidium by Eric Cornell and Carl Wieman at JILA and in sodium clouds by Wolfgang Ketterle at MIT. Like a laser beam, the atoms released from an atomic vapour BEC can form a beam with extremely well-defined energy, direction, and oscillation phase. Such a beam is known as an atom laser.

[1] Some atoms behave as bosons, others as fermions. It all depends on the total angular momentum they have. When two fermions try to occupy the same state, their wavefunctions interfere destructively to make no vibration at all, so two fermions can never be together in the same state.

Quantum Mechanics and Biology

Johnjoe McFadden Professor of Microbiology,
University of Surrey

It has been more than 50 years since Erwin Schrödinger made the startling proposal that life is based on quantum-mechanical principles. Yet most of his arguments remain valid today. Schrödinger pointed out that classical laws are all statistical, true for collections of billions of atoms or molecules but invalid at the level of individual particles. Life, he argued, was based on the dynamics of individual particles and thereby subject to quantum laws.

This was a remarkable prediction in 1944, when the inside of the cell was generally considered to be an amorphous kind of jelly called protoplasm. But as molecular biologists have delved deeper into the workings of cells they have uncovered structure at all levels. The DNA double helix is only two nanometres (millionths of a millimetre) wide – not far above the atomic scale – and, along with proteins and other cellular constituents, has structure at less than one-tenth this size. Dynamics in these biomolecules is all about the motion of individual particles. Consider the enzymatic rotary engine, F1-ATPase, that makes ATP (the cell's chemical fuel). This tiny biological device is just ten nanometres across. The enzyme sits in the cell membrane where its rotation is driven by the flow of protons through a central pore. But how rotary motion at these scales is converted into chemical energy is a mystery. Quantum dynamics are certainly involved.

Quantum tunnelling is thought to be involved in a wide range of biochemical processes including photosynthesis, respiration, mutation, and protein folding. For

instance, enzymes that establish the proton gradient that drives the F1-ATPase are thought to use electron tunnelling to connect the proton transport process to proteins of the respiratory chain. Indeed both electron and proton tunnelling are considered by many to be the missing ingredients that account for the ability of enzymes to massively accelerate chemical reactions. Quantum tunnelling may also be involved in protein folding, the process by which a protein molecule searches through billions of possible structures to find its active form. And quantum tunnelling may be fundamental to the evolution of life on Earth. Watson and Crick first suggested that DNA base tautomerization – a chemical euphemism for proton tunnelling – within the double helix is responsible for mutations. This mechanism has been further explored by myself and Jim Al-Khalili to propose that quantum coherence may play a role in certain types of mutation.

The living cell is nature's nanotechnology. Just as engineers and physicists working at the nano-scale level must include and exploit quantum mechanics in their models, so evolution over three billion years must have incorporated quantum dynamics. Quantum mechanics is likely to be as fundamental to life as water. Indeed, recent experiments and simulations indicate that protons involved in hydrogen bonding in water are highly delocalized (that is, in a superposition of being in two separated locations). Hydrogen bonding is probably the most fundamental biochemical interaction, involved in DNA base-pairing, enzyme catalysis, protein folding, respiration, and photosynthesis. If quantum delocalization lies at the heart of this phenomenon then it is central to life. Indeed, several researchers (including myself and Paul Davies) have proposed that quantum mechanics may account for that ultimate of biological mysteries: the origin of life itself.

Chapter 10
Into the New Millennium

We have seen the role quantum mechanics has played and the impact it has had in so many areas of physics and chemistry, as well as playing a vital part in many of the technological developments of the last century, such as the laser, the semiconductor, and the nuclear reactor. So it is time, in this final chapter, to return to some basic ideas again, such as superposition, entanglement, and decoherence, to see what role they might play in the technology of the 21st century.

Clever experiments

I wonder what Niels Bohr would have made of some of the incredible experimental developments in the fields of atomic physics and quantum optics over the past ten years. The early quantum pioneers insisted that quantum mechanics should only be applied to a large number (or ensemble) of identical quantum systems for its predictions to be meaningful. They also insisted that there should be a clear demarcation between the microscopic world where the quantum rules applied and the macroscopic world of measuring devices that behaved according to classical physics.

Both these restrictions have now been lifted as experimenters work with individual atoms and photons.

Here is where incredibly precise lasers have found another use: as a tool to manipulate individual atoms. Once atoms are trapped in an ultra-high vacuum using electromagnetic fields, precisely tuned and aimed lasers can be used to cool them down. I should point out that the idea of an atom having a 'temperature' simply means that it is jiggling about; the more agitated an atom is, the higher the temperature. The laser light prompts the atom to lose energy and 'cool down'. Lasers can now routinely be used to cool atoms down to less than a thousandth of a degree above absolute zero. They can then act as optical tweezers to hold them in place, and can finally pump them with precise bursts of energy that send them into quantum superpositions and entangled states.

What is remarkable is that these techniques have only been successfully applied in the final decade of the last century, and papers reporting their success have been filling the pages of *Nature* over the past few years. We are finally able to probe the boundary between the quantum and classical worlds.

To give you an example of how rapidly ideas are changing, physics students have long been taught that in order to 'see' an object it cannot be smaller than the wavelength of the light that is shone on it. This is what prompted the invention of the electron microscope described in the box in the last chapter. Visible light has a wavelength of around half a micron (a thousandth of a millimetre) whereas atoms are more than a thousand times smaller still. So textbooks will have to be rewritten now that researchers can almost routinely trap, look at, track, and generally mess around with individual atoms using laser beams of visible light.

Before I move on to describe how some of these techniques might one day be used to build a device called a quantum computer, let us have a look at a few of these experiments. How can visible light allow us to see something as small as an atom? Surely its wavelength is far too large.

How to track an atom

Physicists are now able to detect a sample of a few thousand atoms by shining light on it, but not in the way you might think, of simply bouncing light off the sample as we could do using a microscope. If the frequency of the light is tuned to ensure that its photons have an energy, through Planck's formula, that corresponds exactly to the energy of an atomic transition, then some of the photons will get absorbed by the atoms. Remember that a transition just means that an electron in the atom jumps to a higher energy level. So with light at this 'resonant' frequency, we see a reduction in the overall number of photons due to those few thousand brave photons that sacrifice themselves to the atoms. This is how we know the atoms are there.

A related and highly ingenious idea allows the detection of a single atom. Instead of using laser light tuned to the resonant frequency, we rely on what happens to light that is not quite at the right frequency to be absorbed. Single atoms are first trapped and cooled down, then they are injected one at a time into a tiny device called an optical cavity that is a fraction of a millimetre long and has highly reflective walls. Light from a very weak laser is shone into the cavity so that, as well as the atom, there is on average just one photon bouncing around at any given time! Every time the photon meets the atom it is slowed down ever so slightly as it travels 'through' the atom[1] (in the same way that light will slow down when passing through water or glass). This causes a slight change in the photon's wavefunction that gradually builds up as it passes through the atom thousands of times until the effect is measurable.

Physicists at the Max Planck Institute for Quantum Optics in Germany have used this technique to track the trajectory of an atom

[1] Technically, we say that there is a shift in the photon's phase. The notion of a phase is connected with wavelike behaviour – as in the terms 'in phase' and 'out of phase' – and yet can be applied to a single photon of light, which is described by its wavefunction – a quantity that carries phase information.

as it roams around within the cavity. Of course what is going on here amounts to a continuous observation of the atom, so it behaves as a classical particle at all times.

Watching decoherence at work

A number of experiments in the field of quantum optics have really hit the scientific headlines in the last few years. For three-quarters of a century, theoretical physicists and philosophers have debated such fundamental ideas as where the boundary between the quantum and classical worlds lies by using thought experiments and arguments based on differing views over the interpretation of quantum mechanics. Now, finally, these ideas can be put to the test in the laboratory.

In Chapter 5, I described how decoherence has solved one of the two fundamental issues concerning the measurement

The 1996 experiment in Boulder, Colorado which created the first 'Schrödinger cat' state of an atom, by first trapping the atom in a force field and then slowing it down by laser cooling. Two further lasers were then used to 'force' it into a superposition of being in two places at once.

problem – namely, why we never see cats that are both dead and alive at the same time. The phenomenon of decoherence shows, as we should rightly expect, that there is no sharp dividing line between the micro and the macro worlds, but rather that the interference effects of superpositions disappear increasingly quickly with increasing complexity of a quantum system. So they get destroyed extremely rapidly as soon as a quantum system comes into contact with its macroscopic environment. The trick is therefore to study 'mesoscopic' systems, which are somewhere in between, in the hope that decoherence can be caught in the act.

In May 1996, a group of experimenters at the National Institute of Standards and Technology (NIST) in Boulder, Colorado created what they called a 'Schrödinger cat'-like state, where the 'cat' in question was a single atom. After first trapping the atom and laser cooling it down to as close to absolute zero as possible without violating the uncertainty principle,[2] it was then zapped by a sequence of controlled laser pulses that forced it into a superposition of two different quantum states based on the energy of the atom's outer electrons.

By itself, this is not very remarkable. Atoms are often in superpositions. What was clever was the fact that the lasers got the different states of the atom quantum entangled with states of its motion, so that it was in a superposition of moving in two directions at once too. The atom would oscillate to and fro inside its trap, with the two states moving completely out of phase. Their furthest separation would be almost a thousand times the diameter of the atom. Note that when I say 'they' I am talking about the two parts of the wavefunction of a single atom.

Ah, but I know that wavefunctions are not localized in space so this kind of behaviour should not sound so strange. Even so, each

[2] At absolute zero an atom would have to be stationary (having zero momentum) and fixed at a precise location. But then we would know both its position and momentum exactly, in violation of the uncertainty principle. Atoms will therefore always have a small amount of energy known as their zero point energy and we can never quite get down to absolute zero.

quantum
mouse atom

microwaves force the
atom into a superposition

photons in cavity are forced
into superposition by atom
passing through

quantum
cat atom

measurement

The 1996 Paris experiment that first confirmed decoherence to be a real physical
process. By studying the states of two atoms (the quantum cat and the quantum
mouse) passed through a cavity containing an electromagnetic field in a superposition,
they could measure how quickly the field decohered.

oscillating piece of the wavefunction has a spread of just one-tenth
of the maximum separation between the two pieces. This is why such
a situation differs from the usual two-slit experiment. After all, an
atom that has gone through the two slits is also in a superposition
of being in two places at once. But in that case the two pieces of the
wavefunction spread out as soon as they emerge from the slits and
overlap, and hence interfere. Here each piece of the wavefunction
remains localized and does not spread out in space. When the two
pieces are at their furthest separation there is very little overlap.

Once such mesoscopic Schrödinger cat states could be made,
the next step was to use them to probe the nature of decoherence.
By December of 1996, the first experiment to achieve this had been
carried out by a team led by Serge Haroche working at the Ecole
Normale Supérieure (ENS) in Paris. Instead of just putting an atom
into a superposition, they managed to entangle the state of the
atom with an electromagnetic field consisting of just a few photons
trapped inside a cavity. In this way, the electromagnetic field was

also forced into a superposition, that of oscillating with two different phases at once.

What they did next was to measure how long the electromagnetic field could hang on to this quantum superposition. An interaction with its environment would take place as soon as a single photon leaked out of the cavity to betray the quantum state of the field to the outside world. To measure how quickly this happened they sent in a second atom, which they referred to as a quantum mouse,[3] that would itself become entangled with the quantum state of the field. This led to interference between different parts of the second atom's wavefunction that they could measure. By changing the time interval before the Schrödinger mouse atom was sent in they were able to witness decoherence in action. The speed with which the superposition of the electromagnetic field is lost depends on just how 'out of phase' the two components are. Typically, this decoherence time can be stretched to a tenth of a millisecond. Here at last was dramatic proof that decoherence was real.

More recently, there has been much interest in the prospect of what is called environment engineering, which basically involves warding off decoherence by maintaining the quantum superpositions of trapped atoms for as long as possible. So how is this done? Yet again, using lasers. Only now the job is much harder and requires many lasers working together. So once you include the lasers that trap the atom, the lasers that cool it and the lasers that set it into a superposition, we're talking big business for the laser industry!

Record-breaking entanglement

Catching decoherence in action and controlling it has important applications in the emerging fields of quantum cryptography, quantum computing, and quantum teleportation, which I will discuss in turn in this chapter. But before I do, it is worth briefly mentioning

[3] Since it would be checking on the state of the quantum cat.

recent progress in harnessing another of the fundamental properties of the wavefunction – entanglement.

We will see when I come to discuss quantum computing that in order to make real use of quantum superpositions we need to entangle many quantum states together. During the 1990s various groups were successfully entangling two or three atoms or photons together, but it was not easy. An interaction with the environment by any of the entangled particles would represent a measurement and bring about the collapse of the delicate superpositions like a house of cards. Then in March 2000, the NIST group reported in the pages of *Nature*[4] the successful application of a new technique for entangling a string of four trapped atoms. Each atom was put into a superposition and the four were then entangled together. The method, they claim, is scalable up to much larger numbers of particles.

Another breakthrough was reported in September 2001 by a group in Aarhus in Denmark who managed to entangle the quantum states of two macroscopic objects: samples of caesium gas, each containing trillions of atoms! This was maintained for almost a whole millisecond. OK, don't laugh, I know a millisecond is not very long, but this is still very impressive. You see, if each sample had been in a superposition of two states, in each of which all the atoms were doing the same thing – all in one energy state or all spinning the same way – then just one atom leaking out would betray the state of the whole sample, collapsing the superposition. This would involve a decoherence time of less than a femtosecond.[5] However, they managed to maintain the entangled state for a trillion times longer!

To achieve this, they didn't go for what is called a maximal entanglement, where all the atoms in each sample are doing the same thing. Instead, they put each of the two samples of atoms in a

[4] C.A. Sackett et al., *Nature*, vol. 404 (16 March 2000), p. 256.

[5] Take your pick from these definitions: A femtosecond is 10^{-15} seconds; or a billionth of a microsecond; or a millionth of a billionth of a second; or… well, it's very short anyway.

superposition of two states, in each of which slightly more than half the atoms were spinning one way, with the rest spinning the other way.

This way, if an atom leaked out and betrayed its spin state, it would not be enough to collapse the wavefunction of the whole sample into one or other state since the spin state of that atom might have belonged to either state of the whole sample. Thus, the loss of coherence in the state of a single atom that might escape causes only very slight damage to the overall superposition. That is, detecting the state of one atom does not amount to measuring the state of the whole sample.

Quantum cryptography

The above techniques are more than just neat ways of highlighting the weirder aspects of quantum mechanics. They have a practical purpose too: to help realize the dream of one day building a quantum computer. But there has already been one successful application of entanglement. It is called quantum cryptography.

I will first outline what is involved in classical cryptography. If you have ever worried about how safe it is to give out your credit card number when shopping online, don't (worry, that is). It is extremely secure, for the time being at least. Over the years, mathematicians have searched for ways of allowing two parties to exchange information in complete secrecy. The standard way of doing this is to send a coded message and hope that an eavesdropper won't be able to crack it. There are a number of neat tricks for ensuring the security of encrypted messages, such as the 'public key' system. The simplest form of this idea is based on the following example. If I wish to receive a secret message from you I send you an empty, open, impregnable box and an open padlock, for which I own the only key. You put the message in the box and lock it with the padlock, then send it back to me. The padlock is such that it can only be opened with my key.

Systems such as this depend in practice on the idea that certain

mathematical operations are easier to perform in one direction rather than the other, such as multiplication and factorization. If I told you that x times y equals 37523, how long would it take you to factorize this and give me the values of x and y? But if I gave you the reverse problem of 239 times 157 equals z and asked you to work out z I am sure you could give me the answer much more quickly. The most popular method of public key encryption relies on the difficulty of factorizing very large numbers. It takes even the most powerful computers a very long time to do this. A thousand digit number would, for instance, take much longer to factorize than the age of the Universe, even on the world's most powerful computer!

However, if we were ever to succeed in building a quantum computer then we might have a way of factorizing numbers far more quickly. If so, the security of current public key cryptosystems would be compromised very quickly. And even without quantum computers, we cannot rule out advances in mathematics leading to the discovery of an algorithm for factorizing large numbers. Thankfully, there is another type of cryptography that is completely foolproof, and it also relies on quantum mechanics.

The basic idea behind quantum cryptography is that it allows for the exchange of the cryptographic 'key' between two remote parties – known in the literature as Alice (the sender) and Bob (the receiver) – in absolute security that is guaranteed by the laws of physics. This key is what allows the sender to encode and receiver to decode the cipher text. So quantum cryptography should more correctly be called quantum key distribution.

There are two techniques so far developed. Both rely on the fact that, according to quantum mechanics, any attempt by an eavesdropper to intercept the key will involve a measurement of some kind, and this will unavoidably disturb the system and alert the sender and receiver. The first is the Bennett-Brassard protocol, named after its inventors in 1984, and relies on the idea of Alice and Bob measuring and exchanging photons. Certain of their properties, such as their polarization, can then be converted to binary 0's and 1's to make up the key. Without going into further technical detail,

this method relies on quantum superposition and the uncertainty principle.

In the early 1990s, Artur Ekert discovered a second protocol that relies on the ideas of nonlocality and entanglement. Here, Bob sends to Alice one of a pair of entangled photons, which she measures in some way and sends back to him. He then carries out a measurement on the combined state from which he can determine the type of measurement that Alice carried out. His knowledge of a series of such measurements by Alice will make up the key. Any attempt by the eavesdropper to intercept the photon will affect its partner and alert Bob.

Moore's Law

I don't know if I still have it up in the loft somewhere, but it is now over twenty years since I purchased my very first programmable computer. It was a Sinclair ZX81[6] with a processor frequency of 3 MHz and one kilobyte of memory. I did add extra memory by attaching a 16K RAM card to the back, which at least allowed me to write more than one screen's worth of code before I filled up its memory. But any slight nudge and the physical connection, supplemented by a generous supply of Blu Tack, would come loose and wipe out everything I had typed in. In any case, I could never use it for more than writing short programs to work out stuff for my college lab reports that would take me marginally longer on a pocket calculator. In contrast, the laptop I work on today is about the same size but has a processor speed of 1000 MHz (three hundred times faster) and a disk space of 15 Gigabytes (a million times more), and it's already over a year old, so no longer state-of-the-art.

In 1965, the co-founder of Intel, Gordon Moore, prophesied that computer power would, for the foreseeable future, double every

[6] It was marketed in North America by Timex as the TS-1000. In the UK it was the first computer to be available for less than a hundred pounds.

eighteen months. His prediction, now called Moore's Law, has proved to be remarkably accurate, and we are all aware of the need to upgrade our PCs regularly to keep up with ever-more sophisticated software. But how long can things go on for? Do we know when Moore's Law is likely to break down?

It turns out that we do have a good idea when this will happen, if we continue with today's technology. I mentioned in the last chapter that lasers are used to carve out the tiny patterns of an integrated circuit on the surface of a silicon chip. The increase in computing power depends on the continual miniaturization of this process. With current technological advances this should continue unabated for roughly another twenty years as the wavelength of the laser light used is made shorter and shorter. However, it has been suggested that Moore's Law may come to an abrupt end in five to ten year's time due to an unavoidable problem with increasing miniaturization known as thermal noise. But even without this, we will eventually hit what is called the 'point one' barrier. This is when the wavelength of the laser beam is such that each transistor on the microchip will be just 0.1 micron across. This means that about one thousand, etched side by side onto the chip, could fit within the width of a single human hair. By then, the lasers' wavelength will be down in the ultraviolet range and we will need to find a different technique to create even tinier circuit images on the silicon's surface, such as X-rays or even electron beams, which have much shorter wavelengths. Another possibility, as an alternative to further miniaturization of chips, would be to replace silicon with gallium arsenide. This semiconducting material has an atomic structure that would make the circuits on its surface conduct electricity much faster.

Ultimately, we come up against another barrier called the Rayleigh criterion, which states that the minimum resolvable features on the chip can be no smaller than half the wavelength of the beam. Once we reach this stage we must start to take quantum mechanical effects into account. One possibility being studied for instance is to exploit quantum entanglement of photons in the laser beam to beat this limit, but probably not by much.

Once we get down to molecular sizes, roughly by the year 2020, the age of the semiconductor chip will have come to an end. Researchers are looking elsewhere, and there are already two areas that are showing promise. The first is to use biolomolecular computers, which rely on the fact that the DNA molecules can hold an incredible amount of information and so may one day be used to build molecular logic circuits. The other possibility is to use quantum transistors that rely on the properties of single electrons artificially trapped in 'quantum wells' not much bigger than an atom. By altering the voltage slightly across these wells we can control how the electrons behave in a way similar to the action of a transistor.

All the above advances are more than pure speculation and should see computing power continue to grow during our lifetimes. However, an increasing number of quantum physicists are working on really putting quantum weirdness to the test. They are optimistic that at some time during this century they will be able to build the ultimate quantum machine: a quantum computer.

Qubits

In the early 1980s, round about the time I purchased my ZX81, the physicist Richard Feynman speculated that it would make sense to solve certain problems, such as simulating the behaviour of a quantum system, using a computer that also behaves quantum mechanically. Such a computer would utilize the concept of superposition to create completely new types of algorithms. The field of research into quantum computers soon took off when, in 1985, Oxford physicist David Deutsch published a pioneering paper that showed how this might be achieved in practice.

What Deutsch had proposed was a blueprint for a 'universal quantum computer', just as Alan Turing had proposed the idea of a universal classical computer half a century earlier. Deutsch's machine would operate according to quantum principles to simulate any physical process. It required a row of quantum systems that

could each exist in a superposition of two states, such as atoms in superpositions of two energy levels. These quantum systems would then be entangled together to create quantum logic gates that would be made to perform certain operations.

The basic idea is that of the 'quantum bit' or qubit. In a normal digital computer, the basic component is the 'bit', a switch that can be in either of two positions: off or on. These are denoted by the binary symbols of 0 and 1. However, if a quantum system, such as an atom, is used then it could exist in the two states at once. A qubit can thus be both off and on at the same time, just as long as it can be kept isolated from its environment.

Of course a single qubit is not very useful. But if we entangle two or more qubits we can start to see the power of such a set-up. Consider the information content of three classical bits. Each can be either 0 or 1 and so there are eight different combinations of the three (000, 001, 010, 100, 011, 101, 110, 111). But just three entangled qubits allow us to store all eight combinations at once! Each of the three digits is both a 1 and a 0 at the same time.

Adding a fourth qubit would give us 16 combinations and a fifth, 32 and so on. The amount of information stored increases exponentially (as 2^N, where N is the number of qubits). Now imagine carrying out operations in the same way that we would with classical bits. We would be able to perform 2^N computations at once, the ultimate in parallel processing. Certain problems that might take a normal supercomputer years to solve could be cracked in a fraction of a second.

So what could a quantum computer do?

This all sounds very well, but how exactly can we make use of it to solve a real problem? After all, if classical computers are getting faster and can be linked up via the Internet to work in parallel, won't we be able to eventually match this sort of performance anyway? The answer to these questions was discovered in 1994 when Peter Shor, working at AT&T's Bell Labs in New Jersey, devised the very first

Quantum Computing

Andrew Steane Lecturer in Physics, University of Oxford

Today's computers, for all their marvels, operate on the same fundamental principle as the mechanical devices dreamed up by Charles Babbage in the 19th century and later formalized by Alan Turing: One stable state of the machine represents one number. Even seemingly nonstandard computation models, such as the one based on DNA, share this basic principle. What else could they do? Yet the laws describing the natural world are the subtle laws of quantum physics, and they invite us to think differently about computing. Quantum physics offers powerful methods to manipulate information which we are only just beginning to understand.

Quantum computing brings together two of the most significant conceptual revolutions of the 20th century: information science and quantum physics. It turns out that they fit together very well, because the language of quantum physics is very much like the language of information: the quantum wavefunction is defined to be a mathematical entity which expresses all the properties of a given physical system, so the wavefunction is like a quantity of information. Physical processes cause it to change, so if we design appropriate processes, then the evolution of the wavefunction will be a form of information processing which we can control. What is significant and exciting about quantum computing is that it turns out this quantum evolution can be used to take very powerful short-cuts in certain computations, which are simply not available to other devices.

The simplest piece of quantum information is the qubit, the quantum cousin of the classical 'bit', a switch which can have two positions, either 0 or 1 – while a qubit can be in a

superposition corresponding to both 0 and 1 at the same time. A single qubit might be stored in a single atom or a single photon of light. The subtlety and the power of quantum computing comes in the next step. If all we could do was manipulate many qubits separately, then the computing achieved would have no advantage in processing power over conventional computers. However, according to quantum physics, qubits can be in entangled states. These are states in which the state of one qubit is closely correlated with the state of another. For example, they might be in a state where they are guaranteed to store the same number (both 0 or both 1) but where neither qubit on its own stores either 0 or 1.

In a quantum computer, entangled states involving many qubits are used. The computer is first prepared with all qubits in a simple state such as 0 (spinning one way – up or down). Then qubits are linked to one another via quantum `logic gates' which cause the spin of one qubit to become entangled with that of another, the details being set by the program which the machine is running.

The process continues, but the program is carefully designed so that when it is finished the qubits are once again disentangled – that is, they are back in simple states such as down or up (0 or 1). The final state of the whole collection gives the outcome of the computation. It can be read out by making measurements on the qubits.

It is at present not clear how best to understand the computational advantage which entanglement offers. One feature is that since any qubit can contribute both 0 and 1 to any given state, each time a single qubit is added to the computer, the number of contributions to the total computer state doubles, leading to truly vast numbers of contributions (e.g. $2^{100} = 1,267$ billion billion billion for just 100 qubits). In some respects, the quantum computer acts like a collection of a vast number of traditional computers, all calculating at once! However, that picture does not capture the whole of what is going on, because the separate strands of any such quantum calculation cannot be completely independent, and also they have

to be brought back together (made to interfere with one another) before a meaningful output can be gained.

In the most famous quantum program, Shor's factoring algorithm, entanglement is used to allow two groups of qubits to store a collection of correlated numbers, where each number in the second group is given by a fixed mathematical operation (such as exponentiation) on the corresponding number in the first group. The quantum programme instructs the computer to manipulate the first group in such a way that, through the wonders of entanglement, a shared property of all the numbers in the second group is revealed, such as whether they are all even, or all multiples of some unknown number to be discovered. This information can be used to identify the prime factors of any integer, which, incidentally, would enable most secret encryption schemes currently employed in the business and political world to be broken.

The race is now on to build the first quantum computer, but entanglement is delicate and so far the sufficiently precise control of large numbers of qubits is beyond our technological capability.

quantum algorithm, a set of instructions to carry out a task that only a quantum computer could do. The task was that of factorizing very large numbers incredibly efficiently, one of the central problems in computer science. It immediately became apparent that quantum computers, if they could be built, would have a major impact on banking and commerce since currently secure public key encryption methods would be rendered useless, which is where quantum cryptography comes in of course.

A few years later, another quantum algorithm was discovered by a colleague of Peter Shor, mathematician Lov Grover. Grover's algorithm allowed a quantum computer to search an unsorted database much more quickly than a conventional search engine ever could. Consider a simple example: if I asked you to find a particular card in a completely shuffled pack, then you have a one in fifty-two chance of picking it first time. You might get lucky of course, or you might be unlucky and after checking and removing card after card find it is the very last one in the pack. After many trials, you would find that the average number of guesses you need is twenty-six (half the pack). A quantum computer using Grover's algorithm would be able to find the card, on average, in about seven tries. The mathematical connection is that, in a database of N items, a classical computer needs N/2 tries, whereas a quantum computer would find it in a number of steps equal to the square root of N.

While Grover's algorithm is not as spectacular as that of his colleague Peter Shor, it really comes into its own in, say, a chess match between a quantum computer and a classical computer, since the former would be able to think moves through billions of times faster.

How to build a quantum computer

There are currently several ways of realizing the dream of building a practical quantum computer. All rely on the idea of manipulating entangled superpositions of atoms, but all ultimately suffer with the

Quantum logic gates

A logic gate is a computing device that performs a simple operation if fed one or more bits of information. The way it does this is based on a field of mathematics known as Boolean algebra after George Boole, who developed it in the 19th century. Complex logic circuits that make up a computer's 'brain' are built up from such gates that receive binary inputs of 0's and 1's then follow simple instructions to do something with them. A transistor acts as a logic gate that carries out such an operation by converting its two inputs, each of which can be 0 or 1, into a single output that is 0 or 1.

There are various kinds of logic gates, such as the 'AND' gate (only when both inputs are 1 will the output be 1) and the 'OR' gate (the output is 1 when either or both inputs are 1). Building up combinations of such simple gates, together with the 'NOT' gate (which takes one input and switches it round: 0 to 1 or 1 to

0) allows us to construct more sophisticated logical operations. Thus, by combining together a couple of NOT gates, a couple of AND gates and an OR gate you have the rudiments of an adding device (called an exclusive OR gate, or 'XOR' gate).

Quantum logic works in the same way, only now we must track all eventualities. Quantum algorithms such as those of Shor and Grover rely on the special order in which logical operations are performed on two or more qubits. However, instead of passing electric currents through semiconductor diodes we now manipulate superpositions of quantum states with lasers or magnetic fields.

The simplest example of a qubit is a single electron (or any particle with a quantum spin that can point either parallel or antiparallel to an applied magnetic field). Applying an additional electromagnetic pulse for the right amount of time can flip the electron's spin over. This is an

example of a quantum NOT gate. Another quantum operation is one that corresponds to only turning the electron's spin over half way. This puts it into a superposition of spinning up (1) and down (0) at the same time. This is referred to as a 'square root NOT' (SRN) operation. With two entangled electrons both starting off spinning the same way, an SRN operation puts them into a superposition of four possible states: 00, 01, 10 and 11. With more qubits, complicated quantum algorithms can be built up. For instance, researchers have succeeded in making the quantum equivalent of an XOR gate, allowing them to perform simple addition.

Once the algorithm has been completed, one of the possible final states is selected and has to be amplified so as to be readable using some macroscopic device. Of course, this is just one of many outstanding problems of implementation that have yet to be solved.

Quantum cloning

Many of the exciting new advances and promising future applications of quantum mechanics, such as quantum cryptography and quantum computing, rely on the relatively new field of quantum information theory. An, as yet, not fully explored subject that will have a bearing on all of these technologies is quantum cloning.

While we are familiar with the advances in genetics and the cloning of animals, and possibly one day humans too, it is important to stress that in these cases the clone is not identical to the original, but rather has only the same genetic blueprint. In quantum mechanics, a clone is in every sense identical to the particle or quantum system it was copied from. In 1982, William Wooters and Wojciech Zurek came up with a simple mathematical proof showing that the perfect cloning of an arbitrary quantum system is impossible. If we already know

what the quantum state is beforehand, of course, we can in principle construct a quantum cloning device that will produce identical copies of it, but no one such cloning device can be used universally on all quantum systems.

To clone an object we must first learn everything about it. This we must do through measurement. Once we have gathered all the necessary information we can use it to construct the clone. But you have seen by now that measuring a quantum system does not allow us to do this – something always gets lost in the act of measurement. This is because we are converting quantum information to classical information. For instance, a photon that is in an arbitrary superposition of different spin or polarization states will, when measured, yield one or other of these states but never the superposition. We therefore cannot know the relative amplitudes of the different parts

of the original superposition, or how they were combined (their phase). This is not good enough for cloning.

Since Wooters and Zurek's proof, researchers have figured out that they are able to, in theory, construct what is called a universal quantum cloner which, while never perfect, can have a well-determined success rate, or 'fidelity'.

Quantum cloning could prove useful if we ever build a quantum computer. Rather than performing sequential operations on a qubit, the qubit can first be cloned several times, providing much more efficient parallel processing with all the clones working at once. Of more immediate interest is to understand how safe quantum cryptography will be if an eavesdropper is able to make even approximate clones of photons used in transmitting messages.

same problem: how to prevent decoherence from taking place and destroying the delicate computation. I will outline here a couple of techniques being currently investigated. One employs the idea of manipulating ultra-cold atoms with lasers, and the other uses NMR (nuclear magnetic resonance).

The first method was one I mentioned earlier when I described the various experiments carried out by the groups in Paris (the ENS group) and in Colorado (the NIST group). For instance, the NIST group has suggested a method along the lines of David Deutsch's original proposal in his 1985 paper. They suggest trapping a string of atoms, about twenty microns apart, inside what is called an ion-trap quantum processor. Pairs of crossed laser beams force each atom (acting as one qubit) into a superposition of two energy states. The atoms are in fact charged ions and so feel each other's electrical repulsion, and are thus in communication with each other and hence collectively in an entangled state. They vibrate to and fro in a controlled way, with their relative motion coupled through the exchange of quanta of vibrational energy.

Another possibility is to use the technique of nuclear magnetic resonance, whereby the spin of atomic nuclei in designer molecules is controlled with a magnetic field, and each nucleus behaves as a minuscule bar magnet. We cannot, of course, keep track of the spin states of individual nuclei, but rather it is the bulk properties of the material, containing a hundred billion trillion molecules, that matter. Each molecule constitutes an NMR quantum processor, with the qubits being the nuclei of the atoms that make up the molecule.

An example of such a molecule is chloroform, which consists of five atoms (one carbon, three chlorine and a hydrogen). Instead of using the common isotope of carbon, carbon-12, whose nucleus has an overall spin of zero, the rare isotope carbon-13 is used. Its nucleus has a spin due to the extra neutron it contains. This nucleus, along with the proton (the nucleus of the hydrogen atom), can have two different spin directions (their spin quantum number is one-half, so they can have spin 'up' or 'down'). By applying a radio-frequency pulse to the proton, it can be put into a superposition of spinning in both directions at

The 2001 world-beating seven-qubit quantum computer from Stanford University and IBM's Almaden Research Centre in California. This molecule, known as a perfluorobutadienyl iron complex, used in an NMR system, was the first to successfully implement Shor's algorithm to factorize a number, albeit the number '15'!

once. The close proximity of the two nuclei and the chemical bonding within the molecule ensures their states are entangled and the superposition of the proton's spin state is transferred to the carbon nucleus.[7]

Finally, I will briefly mention just one of several other current ideas for building a quantum computer. This one, still very much

[7] What happens is that changes in the spin orientation of the nuclei affect the wavefunctions of the orbiting electrons in the two atoms, and hence the binding between them.

The Quantum Brain

There is one final interpretation of quantum mechanics that we have not yet discussed. It deserves a mention for two reasons; the first is that its originator is one of the most respected mathematical physicists of his generation, Roger Penrose. The second is that it may explain the one area of science more mysterious than quantum mechanics: the origin of consciousness.

According to Penrose, superpositions of different quantum states do not collapse due to the act of measurement, the presence of a conscious observer, or even interaction with the environment. Instead, he believes that the process takes place even for an isolated system via a physical process linked with the very nature of spacetime itself. According to Penrose, the 'objective reduction', or collapse, of the wavefunction takes place due to the different spacetime geometries of each state in the superposition. (Thus, if a particle

is in a superposition of being in two locations, the curvature of spacetime will differ according to where the mass of the particle is more likely to be.) Once the difference in geometries builds up to a critical level, as when the particle becomes entangled with its environment, the superposition becomes unstable and collapses into just one of the possible states. Of course, neither Penrose nor anyone else knows the details of this mechanism since we do not yet have a complete theory of quantum gravity.

This interpretation has been applied by Penrose and Stuart Hameroff to explain how consciousness might get switched on within the brain. I should explain first that they appeal to quantum mechanics as they believe that the way we 'think' is fundamentally unlike the way a computer carries out algorithmic processes. This non-computability of conscious thought, they maintain, must require something beyond classical physics – namely quantum physics.

And they believe they have found just the right biological vessel to protect the delicate quantum coherence within the brain from the external environment.

The brain's neurons contain hollow cylindrical polymers called microtubules. These are in turn made up of individual proteins known as tubulin, which can exist in a superposition of two slightly different shapes. Penrose and Hameroff maintain that microtubules have just the right properties for this superposition to be maintained, and spread to surrounding tubulins. A coherent superposition is thus maintained for a significant time, allowing for pre-conscious processes to emerge. Objective reduction of the superposition takes place when Penrose's critical threshold is reached and consciousness is switched on. Of course this would be going on all the time in the brain. Maybe we do not need to build a quantum computer after all; we each carry one around with us in our heads!

in its embryonic phase, relies on the two applications of quantum mechanics I discussed at length in the last chapter: the laser and the microchip. A cloud of atoms that is laser-cooled to just a thousandth of a degree above absolute zero can be made to float above a semiconductor chip via magnetic fields produced by the tiny electric currents flowing through the chip's integrated circuits. The height and speed of the atoms can be controlled as they are guided along the magnetic field contours. In this way, it may be possible to precisely control the entanglement of the atoms' quantum states.

For all possible types of quantum computer, the obstacle at the moment is isolating the delicate superpositions from their environment. The more qubits that are entangled the faster decoherence takes place. However, progress is being made on various fronts. For instance, in the ion-trap processor, the way the ions interact with their environment can be controlled if the properties of the environment are chosen carefully.

Another clever trick is to use something called quantum error correction in order to compensate for decoherence. This builds in a redundancy that means the same information is spread over many qubits. In this way, if the superposition of one of the qubits is affected in some way then the computation can still carry on reliably if the same information is encoded in other unaffected qubits.

Currently, quantum computers with only a few qubits have been designed. It is feasible that in the near future a quantum computer with up to forty qubits could be made. But for a working quantum computer, we would need to entangle thousands of qubits and stave off decoherence long enough for useful computations to be carried out.

Quantum teleportation

The 'transporter' aboard *Star Trek*'s starship Enterprise is familiar to millions of fans as the device that transfers the ship's crew on and off various planets, not by creating their identical replicas but by destroying and recreating them in some unspecified manner. Since

this is science fiction no one is really bothered about the details, although I am sure 'Trekkies' have their explanation.

The general idea of teleportation is to scan an object in such a way that what gets transported is pure information, which is used to reconstruct the original out of the appropriate raw material (the right kind of atoms) at the destination. In this way, the process can take place at the speed of light, as opposed to physically transferring the original object's atoms, which would have to be shifted by slower means. That does not sound like much of an advantage unless vast distances are involved and transportation at the speed of light is necessary.

This is all harmless nonsense you might think. Well, it was until 1993. That year, Charles Bennett and his group of international collaborators showed that, thanks to quantum mechanics, perfect teleportation is theoretically possible, if only at the quantum scale for the time being. Until then, it was felt that teleporting a quantum system was ruled out by the uncertainty principle: we cannot scan a quantum system and gain all possible information about it in order to reconstruct it somewhere else. This is true of course, but the new ingredient that gets round this problem is... entanglement... again.

The basic idea is the following. Let the particle to be teleported from A to B be known as particle x. The device also contains two similar particles, y and z. These two particles are entangled and sent to each of locations A and B. At site A, we now have the original particle we wish to teleport, x, along with particle y. By measuring the two together in a certain way, we gain certain information about x. The rest of x's information is irretrievably lost, since the uncertainty principle does not allow us to know everything there is to know about it.

However, the act of measurement has now also collapsed the entangled state of y and z and primed z, at location B, with information that is related to the information lost from x when it was measured with y.

The information we have gained by scanning x is now transmitted to B via more conventional means and this, along with the information already in z, gives us all the information we need

intact original

approximate copy

send data

original

raw material

'classical' information

entangled pair

original particle

EPR source

identical
teleported
particle

The difference between quantum teleportation and a normal fax transmission.
Top: When an image is faxed, the original sheet remains intact but only an approximate copy is produced.
Bottom: In quantum teleportation, all the information that makes up the original particle is lost but a perfect copy is produced at the other end.

at the new location B. We can now encode this information into a similar particle to x and we will have, essentially, reconstructed x, exactly, at B.

It is important to stress that quantum teleportation does not mean some sort of instantaneous nonlocal transportation, since part of the information needed to reconstruct the quantum system at its destination still has to be transferred classically (that is, no faster than the speed of light). But the beauty of the method is that the rest of the information, that is lost during the scanning/measuring process, and which it was thought the uncertainty principle would deny us, is retrievable again at the destination thanks to the nonlocal correlations of the entangled particles.

I should also stress that once we have all this information, we are doing more than simply creating a replica of the original particle. We are altering its quantum state – through the act of measurement – and recreating it again. We never need to physically transfer the particle itself since the quantum attributes are everything; information is everything. Two quantum particles in the same quantum state are truly identical in a way that is unlike anything in the classical world. So transferring all the information content of a quantum particle amounts to transferring the particle itself.

Of course, scaling this up to teleport humans is another matter entirely. Can you imagine the degree of entanglement we would need to transport the complete quantum state of all the particles in our bodies? Just think of the decoherence headache!

No one can foresee how quickly advances will be made in the future. Some physicists believe we will never succeed in building a quantum computer, while others feel it is a matter of a decade or so away.

Whatever the case, one thing is for sure: we have not seen the last of the quantum.

The future is bright. The future is the Quantum.

Further Reading

There are many books that seek to popularize quantum mechanics and related subjects in modern physics. However, many do not succeed in getting the fundamental ideas across to the non-expert. Here is a personal selection of those that do:

Jim Al-Khalili, *Black Holes, Wormholes and Time Travel* (Institute of Physics Publishing, 1999). It is perhaps not surprising that I recommend this book as the clearest explanation of Einstein's theories of relativity and the nature of space and time ever written!

Julian Barbour, *The End of Time* (Phoenix, 1999). A tough book for the uninitiated but worth the effort. Barbour describes a radical new theory about the nature of time. His suggestion: abandon the concept of time altogether, it is just an illusion. Along the way he describes relativity theory and quantum mechanics and how they might be reconciled.

James T Cushing, *Quantum Mechanics: Historical Contingency and the Copenhagen Hegemony* (University of Chicago Press, 1994). The late Jim Cushing suggests that had David Bohm been around in the early days of quantum mechanics his quite different take on the meaning of the theory may well have been the standard one rather than the Copenhagen view. This is a thought provoking book if a little tough going for the non-expert.

Paul Davies and Julian Brown, *The Ghost in the Atom* (Cambridge University Press, 1986 and Canto, 1993). After a clear exposition of quantum theory in the first chapter, the rest of this book is based on a series of interviews on BBC Radio conducted by Davies with several leading quantum physicists. It provides various unique personal views on the meaning of quantum mechanics. A very enjoyable read.

Brian Greene, *The Elegant Universe* (Vintage, 2000). Award winning and comprehensive book on the search for a theory of everything. Greene is one of the leading experts on string theory and while this is a long book it is well worth it. Beginning with relativity theory and quantum mechanics, the reader is led all the way to ten and eleven dimensional spacetime.

John Gribbin, *Schrödinger's Kittens and the Search for Reality* (Weidenfeld & Nicolson, 1995). Gribbin is an established master in the game of demystifying quantum mechanics. This book is a follow-up to his classic *In Search of Schrödinger's Cat* (which introduced a whole generation to the mysteries of quantum mechanics in the 1980s). Here, Gribbin is critical of the standard Copenhagen view and aligns himself instead with another interpretation, You'll have to read the book to find out which one!

Tony Hey and Patrick Walters, *The Quantum Universe* (Cambridge University Press, 1987). A beautifully illustrated book on the principles of quantum mechanics and the many applications it has in our daily lives, from electronics to astronomy.

Michia Kaku, *Visions* (Oxford University Press, 1998). A wonderfully inspiring book that looks ahead to how 21st Century science is likely to change our lives, from genetic engineering to artificial intelligence, quantum computers and much more.

David Lindley, *Where Does the Weirdness Go?* (BasicBooks, 1996). This non-technical book does a remarkably good job of demystifying the concepts behind quantum mechanics; and the author puts a strong case in support of the standard (Copenhagen) interpretation of quantum mechanics.

Ray Mackintosh, Jim Al-Khalili, Björn Jonson and Teresa Peña, *Nucleus: A Trip into The Heart of Matter* (Canopus, 2001). Probably the only coffee-table glossy on nuclear physics, this book offers many insights into the nature of atoms and their nuclei. It describes how nuclear physics was born and the role quantum mechanics has played in its development as well as the many applications it has today, from nuclear medicine to explaining how the sun shines.

J.P. McEvoy and Oscar Zarate, *Introducing Quantum Theory* (Totem Books, USA, 1996 and Icon Books, UK, 1999). A wonderfully illustrated step-by-step tour of the history of quantum theory as it unfolded through the work of the key players throughout the 20th Century. The illustrator Oscar Zarate won the Will Eisner Prize for best graphic novel, but this enjoyable and easy going romp is no fiction.

N. David Mermin, *Boojums All The Way Through* (Cambridge University Press, 1990). A collection of essays by a leading quantum physicist. This is quite simply the best place to read about Bell's theory and nonlocality in quantum mechanics, but there are numerous other fascinating and amusing nuggets from this renowned science communicator.

Roger Penrose, *The Emperor's New Mind* (Oxford University Press, 1989). This book created quite a stir when it came out and led to a number of international conferences to discuss Penrose's ideas on the nature of consciousness and artificial intelligence. It has a very nice section on quantum mechanics and describes a possible route towards a theory of quantum gravity.

Lee Smolin, *Three Roads to Quantum Gravity* (Weidenfeld & Nicolson, 2000). A practitioner's view of the way ahead in the search for a theory of quantum gravity.

Daniel F. Styer, *The Strange World of Quantum Mechanics* (Cambridge University Press, 2000). Dan Styer gives a clear description of the meaning of the Einstein-Podolsky-Rosen experiment and Bell's theorem. While tougher going than many, this book is worth the effort if you are prepared to work at it.

And if this lot still isn't enough, you can always check out the following books, many of them popular classics and best-sellers: J.C. Polkinghorne (*The Quantum World*), Frank Close (*The Cosmic Onion*), George Gamow, (*Mr Tompkins in Paperback*), R.P. Feynman (*QED: The Strange Theory of Light and Matter*), Paul Davies (*Superforce*), David Deutsch (*The Fabric of Reality*).

Picture Credits

David Angel 3, 4, 5, 8, 10, 14, 17, 28, 30, 32, 36, 39, 49, 52, 56, 60, 68, 70, 71, 73, 88, 90, 99, 109, 113, 117, 122, 129, 135, 138, 149, 150, 151, 154, 156, 162, 167, 169, 174, 176, 185, 193, 194, 205, 211, 222, 227, 230, 237, 239, 262; Tim Oliver 257.

Acknowledgements

A number of friends and colleagues have helped me greatly in writing this book. First and foremost I must thank my wife Julie and children, David and Kate, for their support and for being so understanding during the many evenings and weekends over the past couple of years when I had to shut myself away with my PC. I also owe a huge debt of gratitude to the following people for contributing essays, for reading and commenting on some or all of the manuscript, for offering advice and suggestions as well as correcting the many mistakes. So, in alphabetical order, they are: Jeremy Alam, Julie Al-Khalili, Nazar Al-Khalili, Reya Al-Khalili, David Angel, Marcus Arndt, Michael Berry, Frank Close, Paul Davies, Jason Deacon, Chris Dewdney, Gregers Hansen, Deen Harman, Ed Hinds, Ron Johnson, Greg Knowles, Johnjoe MacFadden, Ray Mackintosh, Abdel-Aziz Matani, Gareth Mitchell, Andrew Steane, Paul Stevenson, Ian Thompson, Patrick Walsh, Richard Wilson and Anton Zeilinger. Naturally, any remaining errors are my responsibility alone. Finally, special thanks go to my editor at Weidenfeld & Nicolson, Nic Cheetham, for all his help.

Index